T0189900

BIOLOGY AND ECOLOGY OF VENOMOUS MARINE SNAILS

BIOLOGY AND ECOLOGY OF VENOMOUS MARINE SNAILS

Ramasamy Santhanam, PhD

Apple Academic Press Inc. | Apple Academic Press Inc.
3333 Mistwell Crescent | 9 Spinnaker Way
Oakville, ON L6L 0A2 | Waretown, NJ 08758
Canada | USA

©2017 by Apple Academic Press, Inc.

First issued in paperback 2021

Exclusive worldwide distribution by CRC Press, a member of Taylor & Francis Group
No claim to original U.S. Government works

ISBN 13: 978-1-77463-607-7 (pbk)
ISBN 13: 978-1-77188-330-6 (hbk)

Library and Archives Canada Cataloguing in Publication

Santhanam, Ramasamy, 1946-, author
Biology and ecology of venomous marine snails / Ramasamy Santhanam, PhD.

Includes bibliographical references and index.
Contents: Introduction -- Biology and ecology of venomous marine snails -- Profile of venomous marine snails -- Conotoxins -- Envenomation of cone snails -- Therapeutic uses of cone snail venoms.
Issued in print and electronic formats.
ISBN 978-1-77188-330-6 (hardcover).--ISBN 978-1-77188-331-3 (pdf)
1. Conidae. 2. Poisonous shellfish. I. Title.

QL430.5.C75S26 2016   594'.3   C2016-904658-3   C2016-904659-1

Library of Congress Cataloging-in-Publication Data

Title: Biology and ecology of venomous marine snails / Ramasamy Santhanam, PhD.
Description: Waretown, NJ : Apple Academic Press, 2016. | Includes bibliographical references and index.
Identifiers: LCCN 2016030005 (print) | LCCN 2016030894 (ebook) | ISBN 9781771883306 (hardcover : alk. paper) | ISBN 9781771883313 ()
Subjects: LCSH: Conidae. | Poisonous shellfish.
Classification: LCC QL430.5.C75 S26 2016 (print) | LCC QL430.5.C75 (ebook) | DDC 594/.3--dc23
LC record available at https://lccn.loc.gov/2016030005

Apple Academic Press also publishes its books in a variety of electronic formats. Some content that appears in print may not be available in electronic format. For information about Apple Academic Press products, visit our website at **www.appleacademicpress.com** and the CRC Press website at **www.crcpress.com**

Biology and Ecology of Marine Life Book Series
Series Author: Ramasamy Santhanam, PhD

- Biology and Ecology of Toxic Pufferfish
- Biology and Ecology of Venomous Marine Snails
- Biology and Ecology of Venomous Stingrays
- Biology and Culture of Portunid Crabs of the World Seas
- Biology and Ecology of Edible Marine Gastropod Molluscs
- Biology and Ecology of Edible Marine Revalve Molluscs

CONTENTS

LIST OF ABBREVIATIONS

CR	critically endangered
CTx	conotoxin
DD	data deficient
EN	endangered
FDA	Food and Drug Administration
GFEMN	Global Fisheries Ecosystem Management Network
GTX II	geographutoxin II
HD	Huntington's disease
IUCN	International Union for Conservation of Nature
MCAo	middle cerebral artery occlusion
NT	near threatened
PPlase	peptidylprolyl cis-trans isomerise
VU	vulnerable

PREFACE

Marine snails form the dominant component of molluscan faunas through-out the world's oceans. The superfamily Conoidea includes chiefly the cone snails (Family: Conidae), turrid snails (Family: Turridae), and Auger snails (Family: Terebridae). Though all these snails are venomous, the cone snails assume greater significance owing to their greater diversity and toxins. There are about 600 different species of cone snails found distributed in warm and tropical seas and oceans worldwide, and the greatest diversity is seen in the Western Indo-Pacific Region. Among these conus species, about 20 species, which are larger and prey on small bottom-dwelling fish, have been reported to be deadly poisonous to humans. Cone snail venoms, such as conotoxins and conopeptides, show great promise as a source of new, medically important substances. The synthetic version of the conopeptide called ziconotide has been approved as a medication in the United States by the FDA (Food and Drug Administration) and is in current use as an analgesic (pain reliever). Conantokins, which are a family of conopeptides found in cone snail venom, serve as "sleeper peptides," and these peptides work by a mechanism that may be helpful for people with epilepsy.

Although several books are available on the poisonous marine life, there is no comprehensive and exclusive publication on the marine cone snails. The chapters of this book include descriptions of about 300 species of hazardous marine snails along with their biological and ecological characteristics; characteristics of conotoxins; cone snail injuries, their treatment, and prevention measures; and therapeutic values of conotoxins. It is hoped that the present publication written in an easy-to-read style with neat illustrations will be of great use for the students and researchers of disciplines such as Fisheries Science, Marine Biology and Zoology, as well as besides serve as a standard reference for all the libraries of colleges and universities.

I am highly indebted to Dr. S. Ajmal Khan, Emeritus Professor, Centre of Advanced Study in Marine Biology, Annamalai University,

India for his valuable suggestions. I sincerely thank all my international friends who were very kind enough to collect and send the different species of molluscan shells for the present purpose. I also thank Mrs. Albin Panimalar Ramesh, for her help in photography and secretarial assistance. Suggestions from the users are welcome.

—Ramasamy Santhanam

ABOUT THE AUTHOR

Ramasamy Santhanam, PhD
Former Dean, Fisheries College and Research Institute, Tamil Nadu Veterinary and Animal Sciences University, India

Dr. Ramasamy Santhanam is the former dean of the Fisheries College and Research Institute, Tamil Nadu Veterinary and Animal Sciences University, India. His fields of specialization are marine biology and fisheries environment. Presently he is serving as a resource person for various universities of India. He has also served as an expert for the Environment Management Capacity Building, a World Bank-aided project of the Department of Ocean Development, India. He has been a member of the American Fisheries Society, United States; World Aquaculture Society, United States; Global Fisheries Ecosystem Management Network (GFEMN), United States; and the International Union for Conservation of Nature's (IUCN) Commission on Ecosystem Management, Switzerland. To his credit, Dr. Santhanam has 15 books on fisheries science and 70 research papers.

CHAPTER 1

INTRODUCTION

Marine snails form the dominant component of molluscan faunas throughout the world's oceans. Although families such as the cowries, cone snails, and murex snails may be the best known due to their attractive shells and often-bright colors, large numbers of ecologically important species are either drab, or small to microscopic in size. The superfamily Conoidea includes chiefly the cone snails (Family: Conidae), turrid snails (Family: Turridae), and Auger snails (Family: Terebridae). Though all these snails are venomous, the cone snails assume greater significance owing to their greater diversity and toxins. There are about 600 different species of cone snails, all of which are poisonous (Hazardous Marine Life—http://www.diversalertnetwork.org/health/hazardous-marine-life/cone-snails). Though larger cone species which prey on small bottom-dwelling fish (piscivores) of cone snails are potentially dangerous, the sting of (mollusc eating-molluscivores and worm eating-vermivores) smaller species are likely to be no worse than a bee sting. Most of the cone snails that hunt and eat marine worms are probably not a real risk to humans, with the possible exception of larger species such as *Conus vexillum* or *C.quercinus* (Anderson and Bokor, 2012; http://diogenes.hubpages.com/hub/The-Conus-Cone-Shells-Beautiful-and-Deadly; http://www.coneshell.net/Pages/pa_cones_venom.htm).

Cone shells of the tropical species are large and attractive, and have received the attention of conchologists for hundreds of years. These species live on coral reefs where they occur under coral slabs or in sand or on the exposed reef surface; or on intertidal sand flats; or subtidally on sand and rubble. Fewer species occur in temperate zones, where they occur intertidally and in deeper water. Intensive research has already been made in cone snails and nothing much is known about the biology and ecology of the other two families.

About 20 species of cone snails like *C. geographus, C. textile, C. striatus, C. gloriamaris, C. tulipa, C. regius* and *C. ermineus, C. aulicus, C. catus, C. imperialis, C. litteratus, C. lividus, C. magus, C. marmoreus, C. nanus, C. obscurus, C. omaria, C. pennaceus, C. pulicarius, C. quercinus* and *C. pennaceus* have been reported to be deadly causing human fatalities. Among turrids all species of *Gemmula* have been reported to be potentially dangerous. Injuries typically occur when the animal is handled. Cone snails administer stings by extending a long flexible tube called a proboscis and then firing a venomous, harpoon-like tooth (radula). A cone snail sting can cause mild to moderate pain, and the area may develop other signs of acute inflammatory reaction such as redness and swelling. Conus toxins affect the nervous system and are capable of causing paralysis, possibly leading to respiratory failure and death.

With regard to toxin types, cone snails yield conotoxins and the other two groups viz. turrids and terebrids yield turritoxins and teretoxins (auger toxins), respectively. While information on toxin types is available for a few species of turrids and terebrds, conotoxins have been identified from more than 100 species. The venom of conus is comprised of a plethora of small peptide neurotoxins, termed conotoxins, which are largely used by these species to subdue prey and for self-defense. These venoms are remarkably diverse as the venom of a single snail may contain upwards of 50 distinct conotoxins and venom composition differs dramatically among species (Olivera et al., 1999). Conotoxins are encoded by large gene families (e.g., A-, I-, M-, O-, P- and T-superfamilies) and so a variety of mechanisms may contribute to the differentiation of venoms among *Conus* species. These conotoxins can be effective in minute quantities, interrupt the transmission of signals in nerve paths in a highly selective manner, and are thus able to block the transmission of pain very well. Consequently, these toxins are of great interest for developing analgesics for chronically ill or terminal cancer patients for whom other medications can no longer be used.

Cone snails may therefore represent an underexplored reservoir of novel actinomycetes of potential interest for drug discovery. Scientists began studying about the cone snail venom in the 1990s only. The synthetic version of the conopeptide called ziconotide (Prialt) has been approved as a medication in the United States by the FDA (Food and Drug Administration) and is in current use as an analgesic (pain reliever). Conantokins, which are a family of conopeptides found in cone snail venom serve as "sleeper peptides" and these peptides work by a mechanism that may be helpful for people with epilepsy (http://animals.pawnation.com/north-american-poisonous-cone-snails-4473.html). Many other peptides produced by the cone snails show prospects for being potent pharmaceuticals, such as AVC1, isolated from the Australian species, the Queen Victoria cone, *Conus victoriae*. This has proved very effective in treating postsurgical and neuropathic pain, even accelerating recovery from nerve injury. Other drugs are in clinical and preclinical trials, such as compounds of the toxin that may be used in the treatment of Alzheimer's disease, Parkinson's disease, depression, and epilepsy.

It is time to investigate how these predatory and defensive venoms are produced and regulated, and use these findings to target those toxins with direct therapeutic potential. (http://www.imb.uq.edu.au/cone-snails-have-multiple-venoms).

KEYWORDS

- **Alzheimer's disease**
- **auger snails**
- **cone snails**
- **Parkinson's disease**
- **turrid snails**
- **ziconotide**

CHAPTER 2

BIOLOGY AND ECOLOGY OF VENOMOUS MARINE SNAILS

CONTENTS

2.1   INTRODUCTION

Conidae is a favorite family of shell collectors. Cones are commercially important in the area and are actively collected for shell trade. Living cones must be handled with great care, as their bites may be painful or even occasionally fatal to humans. Due to the temperature sensitivity of the venom, however, cones are edible without danger after cooking. They are known to be locally used as food in the Indo-West Pacific. It is estimated that about 600 species of cone snails are found distributed in the world oceans. All these species have been to be venomous, producing poisonous substances known as conotoxins that are capable of paralyzing prey and that in some instances are powerful enough to paralyze and kill a human. The use of cone snails is presently limited to the their shells which are valued for ornamentation purposes. However, research is on to isolate several pharmaceutical compounds of therapeutic value from their venoms.

2.2   GEOGRAPHIC DISTRIBUTION AND HABITATS

There are over 600 different species of cone snails. This family is typically found in warm and tropical seas and oceans worldwide, and reaches its greatest diversity in the Western Indo-Pacific Region. However, some species of Conus are adapted to temperate environments, such as the Cape coast of South Africa, the Mediterranean, or the cool waters of southern California (*Conus californicus*), and are endemic to these areas. This genus is found in all tropical and subtropical seas. They are most common in intertidal and shallow sublittoral zones, but also occurring deeper on the continental shelf and slope to a depth of about 600 m. When living on sand, these snails bury themselves with only the siphon protruding from the surface. Many tropical cone snails live in or near coral reefs. Some species are found under rocks in the lower intertidal and shallow subtidal zones (http://www.aquariumof-pacific.org/onlinelearningcenter/species/cone_snails_general_description).

2.3   PHYSICAL CHARACTERISTICS OF CONE SNAILS

Shells of all cone snails are typically cone shaped. They are wide at one end and narrow at the base. All have spires (whorls above the body whorl) of varying heights at the wide end. A whorl is a full turn of the shell. Shell bodies (whorls) can be dull to very shiny, and smooth to lined and bumpy. The shell's aperture is long and narrow and does not have an operculum. A living cone animal has a long foot. The head which is thin has two tentacles, each with an eye about halfway down the outer surface. A flap of tissue called the mantle lines the inside of the shell and is rolled to form a colored siphon that extends beyond the shell and draws water into the gills. The stinging proboscis is normally retracted. Radular teeth are contained in the radular sac; and the size, number, and design of these teeth varies by species. Living shells are well-camouflaged or buried in sand. These living shells are covered with a brown skin-like periostracum, which is the foundation for shell development. The periostracum is often encrusted with coralline algae in species that do not completely bury themselves. Size of cone snails varies widely from 1.3 cm to 21.6 cm in length (Figures 2.1–2.2).

The different parts of a dissected cone snail are given in Figure 2.3.

Apex: the tip of the spire

Anal notch: where the snail excretes solid waste.

Outer lip: the free edge of the mouth. This is where new shell growth would occur.

Inner lip: smooth and usually lacks a color pattern.

Aperture: a long narrow opening where the snail would emerge from.

Base: the base of the shell

FIGURE 2.1 Externals of a cone snail shell.

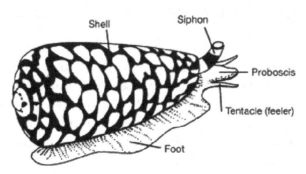

Shell

Siphon

Proboscis

Tentacle (feeler)

Foot

FIGURE 2.2 Organs of a cone snail.

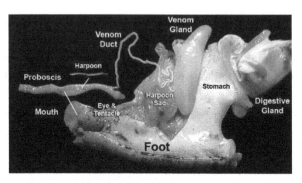

FIGURE 2.3   Anatomy of a cone snail.

2.4   DIET AND FEEDING IN CONE SNAILS

Cone snails are divided into three groups according to their diets viz. Piscivores – fish eaters; Molluscivores – mollusc eaters, and Vermivores – worm eaters.

2.4.1   FISH-EATING SPECIES (PISCIVORES)

Fish-eating species of cones (about 70 of the larger cone shell species) are characterized by shells with wide apertures. *Conus geographus, Conus tulipa, Conus striatus, Conus catus*, and *Conus magus* are exceptionally dangerous and should never be handled under any circumstances (Figure 2.4).

2.4.2   MOLLUSC-EATING SPECIES (MOLLUSCIVORES)

Mollusc-eating species of cones are characterized by prominent tent-markings on the shell. Due to their ability to cause great injury (and occasionally fatality) it is also unwise to ever handle living animals. The larger species such as *Conus textile, Conus aulicus, Conus omaria*, and *Conus marmoreus* should be considered dangerous. The famed rarity *Conus gloriamaris* (a south east Asian and Melanesian species) belongs to this group of cones (Figure 2.5).

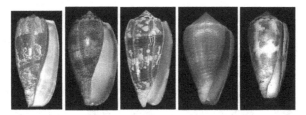

FIGURE 2.4   Fish-eating cone species (Piscivores): [left to right] *Conus geographus, Conus tulipa, Conus striatus, Conus catus,* and *Conus magus.*

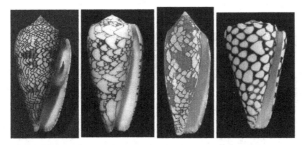

FIGURE 2.5   Mollusc-eating cone species (largest shell: [left to right] *Conus textile, Conus omaria, Conus aulicus,* and *Conus marmoreus.*

2.4.3   WORM-EATING SPECIES (VERMIVORES)

Worm-eating species of cones are characterized by shells with a narrow aperture. The majority of cones species fall into this category. Although none have been responsible for known fatalities larger species, such as *Conus capitaneus, Conus leopardus, Conus vexillum, Conus litteratus* and *Conus miles* should not be handled alive (Figure 2.6).

2.5   CATCHING PREY

Typically a cone snail hunts its prey in two ways (Figure 2.7–2.9). In the first method, cone snails first scent the prey with chemoreceptive cells of the proboscis, gently touch it with the proboscis, and then, lightning fast, sting the prey with the radula (a harpoon-like, poison-tipped tooth which is attached by a thread), and inject the venom. Within seconds, the prey is

FIGURE 2.6 Worm-eating cone species: [left to right] *Conus capitaneus, Conus leopardus, Conus vexillum,* and *Conus miles.*

FIGURE 2.7   A cone snail preying a fish.

FIGURE 2.8   A cone snail preying a worm.

FIGURE 2.9   A cone snail preying a mollusc.

immobilized. The thread is then retracted and the prey engulfed through the expanded proboscis and moved to the stomach to be digested. Recent research findings suggest that conal snails can send the whole schools of fish into hypoglycaemic shock by releasing insulin into the water. These snails use a potent form of insulin to subdue its fish prey. For example, the geographic cone snail (*Conus geographus*) uses its chemicals to cause a plunge in the fish's blood sugar, leaving it sluggish and unable to escape. Subsequently, it can entrap the whole schools of small fish in this way (http://www.sciencedaily.com/releases/2015/01/150119154316.htm).

2.6   RADULA OF CONE SNAILS

The radula of the cone snails varies by their species and diet specialization. Piscivore radula is elongated with a long smooth shaft tipped in long curved barbs. Molluscivore radula has heavy barbs near the base and is serrated over most of the length of the shaft. Vermivore radula is short, broad, and strongly serrated with strong barbs near the middle. Fish-eating cone snails typically hunt at night, when their prey is sleeping, and many secrete chemicals into the water that sedate their prey. This allows the snail to crawl close to its target before releasing a venomous harpoon-like tooth known as a radula. The radula is long, pointed, and adorned with barbs, and each snail keeps about 20 radulae stored in a structure called a radular sac. When within firing range of prey, the snail releases a radula. The barbs at its tip hook into the victim, and the tentacle attached at its opposite end forms a connection to the snail's

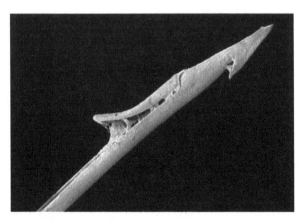

FIGURE 2.10   Radular tooth of a cone snail.

mouth. This allows the snail to hang on to its kill while venom is drawn from a venom gland in the snail's body through a duct in the tooth and delivered into the prey.

2.7   VENOM APPARATUS OF A CONE SNAIL

A cone shell's venom gland is long and shaped like a tube. During gastropods' evolution it has developed from a salivary gland (http://www.molluscs.at/gastropoda/index.html?/gastropoda/sea/conotoxin.html). The structures in the venom apparatus of a cone snail include the organs of venom production and delivery mechanism.

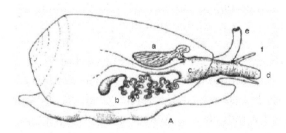

FIGURE 2.11   Venom apparatus of a cone snail. (a) Harpoon sac; (b) Venom gland (Oesophageal gland); (c) Pharynx; (d) Proboscis; (e) Siphon; (f) Eye stalks (tentacles); (A) Foot.

2.8   REPRODUCTION

Although reproduction in cone snails has not been widely studied, it appears that most of these snails have separate sexes and fertilization is internal. Egg capsules of various shapes are laid and attached to substrate. Each capsule contains a varying number of eggs. Two types of hatchlings have been described: the veligers (free-swimming larvae) of variable duration and veliconcha (baby snails).

2.9   BEHAVIOR AND ADAPTATION

Cones are active at night. During the day they rest under stones or coral, or bury themselves in the sea-bed. These slow-moving snails evolved into one of the fastest known hunters in the animal kingdom in their efforts to catch prey. Their average attack lasts only milliseconds. They use stealth and deliver paralyzing venom using stinging harpoons. Cone snail venom is very complicated chemically, varying widely in its makeup from species to species, whether a piscivore, the most toxic, or a vermivore, the least toxic; and with each individual sting or attack. Cone snails are among the most toxic creatures on earth. Over 30 cases of envenomation have been documented worldwide with some fatalities. The venom, which has hundreds of active components, inhibits transmission of neuromuscular signals in the body, initially causing numbing and/or tingling at the site, which spreads to the affected limb, then to the whole body (http://www.aquariumofpacific.org/onlinelearningcenter/species/cone_snails_general_description).

2.10   CONSERVATION OF CONE SNAILS

2.10.1   REGION-WISE OCCURRENCE OF CONE SNAIL SPECIES

Eastern Atlantic: 98 species
Western Atlantic: 113 species
Indo-Pacific: 390 species
Eastern Pacific: 31 species

Source: Peters, et al. (http://journals.plos.org/plosone/article?id=10.1371/journal.pone.0083353).

2.10.2 CONSERVATION STATUS OF CONE SNAILS

A total of 632 species have been assessed for their conservation status and the following table gives the present status.

2.10.3 THREATS

Threats to those Conus species can be classified into four causal groups: (1) pollution, either from proximity to actual or potential petro-chemical spills, or urban and industrial effluent; (2) disturbance to habitat from coastal development either resulting from human population increases, for example, sea defenses, residential and commercial structures, including aquaculture facilities, and port construction, or tourism infrastructure; habitat damage and extensive demersal fishing; (3) shell gathering; and (4) environmental change, for example, elevated sea-surface temperatures.

2.10.4 BIOPROSPECTING

Conus are exceptionally important to biomedical science, although there is dispute about the number of animals taken for their bioactive compounds. To protect their intellectual property, pharmaceutical companies are silent on the issue, but researchers are adamant that volumes are negligible.

TABLE 2.1   Conservation Status of Cone Snails

Category	No. of species %
Critically Endangered (CR)	3–0.5
Endangered (EN)	11–1.7
Vulnerable (VU)	27–4.3
Near Threatened (NT)	26–4.3
Data Deficient (DD)	87–89.5
Total	632–100.0

Source: Peters, et al. (http://journals.plos.org/plosone/article?id=10.1371/journal.pone.0083353).

2.10.5   TEREBRID ECOLOGY AND BEHAVIOR

It is presumed that terebrids also possess three anatomical feeding varieties (Types I, II, and III) only one of which (Type II) possesses the venom apparatus similar to that used by Conus. Questions to be investigated include, how do species that lack the venom duct and radula, which are the main characteristics for capturing prey using venom toxins, feed? Are they still able to capture prey using toxins produced by other glands (e.g., the salivary gland), or did they develop new strategies not based on venom? If so, did the two lineages without the venom ducts develop similar or different strategies? Preliminary work in Conus has demonstrated that peptide toxins are produced in the salivary glands, suggesting it may be possible that the Type I species of terebrid, which lack a venom apparatus, but have salivary glands, could also use toxins to subdue its prey. The delivery of the toxins is not clear as most Type I species do not have a radular or a true proboscis to deliver the toxin to the prey. Type III terebrids have developed an accessory feeding organ that they use to engulf polychaetes and other worms (Puillandre and Holford, 2010).

2.10.6   TEREBRIDAE

The auger shells of the family Terebridae have tall, slender, conical shells with high pointed spires and numerous whorls. The siphonal canals are short, sometimes twisted, with a distinct notch at its base. Most have some sort of axial (radial) ribs. Their common name refers to the shell's resemblance to boring drill bits or screws. They are sand-dwelling carnivores. Many have a venomous barb (similar to cone shells) that is used to stun their prey which consists mainly of worms (http://txmarspecies.tamug.edu/invertfamilydetails.cfm?famnameID=Terebridae).

Unlike the family Conidae, the biology, feeding ecology and phylogenetic relationships of the marine snails in the families Turridae and Terebridae remain poorly understood.

KEYWORDS

- bioprospecting
- molluscivores
- piscivores
- terebrids
- venom apparatus
- vermivores

CHAPTER 3

PROFILE OF VENOMOUS MARINE SNAILS

CONTENTS

3.1   CONE SNAILS (FAMILY: CONIDAE)

The family Conidae has nearly 500–600 species, which have a distinctively similar shape. Characteristic features of this family are: a flat top, conical shell, and a long slit-like aperture lip extending from a very short siphonal opening to nearly the top. Some species have moderate spires, although generally similar. Shell may be smooth or spirally ornamented, and the patterns and colors are extraordinarily varied. All the species of this family are carnivorous and feed on other molluscs (molluscivores), worms (vermivores and small fish (piscivores), which they stun by projecting a venomous harpoon connected to a muscular poison gland. The cone can extend its proboscis lightning fast for a lethal sting and engulf and digest a fish the size of its shell or larger. In some cone species, the venom is powerful enough to be lethal to collectors and sea- goers who are not careful in handling cone snails in live condition. Interestingly, specific components of the toxin complex of certain cone species are currently finding important medical uses.

3.1.1 DEADLY CONE SNAILS

Conus geographus (Linnaeus, 1758)

Apertural view   **Abapertural view**

Class: Gastropoda; **Subclass**: Caenogastropoda

Order: Neogastropoda; **Superfamily**: Conoidea

Family: Conidae

Common Name: Geography cone snail

Geographical Distribution: Tropical Indo-Pacific; Red Sea, in the Indian Ocean along Chagos, Madagascar, Mauritius, Mozambique and Tanzania

Habitat: Sublittoral epipelagic zone; living or fragmented coral reefs; sandy regions within tidal zone

Identifying Features: Shell of this species is broad, thin and cylindrically inflated. Mid-body of the shell is wider and convex and spire is flat-tened, striated and coronated. Ground color of the shell is pink or violaceous white, occasionally reddish. It has a mottled appearance, clouded and coarsely reticulated with chestnut or chocolate with two very irregular bands. It is a piscivore preying on small fish. It releases a venomous cocktail into the water in order to stun its prey. Like the other cone snails, it fires a harpoon-like, venom-tipped modified tooth into its prey. The harpoon is attached to the body by a proboscis, and the prey is pulled inside for ingestion. This species is highly dangerous and it is the most venomous animal in the world. It has the most toxic sting, which is responsible for human fatalities. Live specimens should therefore be handled with extreme caution. There is no antivenom for a cone snail sting, and treatment consists of keeping victims alive until the toxins wear off.

Toxin Type: The venom of this species, among other toxins, contains sodium channel blockers (μ-conotoxins), acetylcholine receptor blockers (α-conotoxins), and calcium channel blockers (ω-conotoxins);

Conantokin-G, a toxin derived from this species showed LD 50 values of 0.001–0.003 mg/kg. The venom of this species is a complex of hundreds of different toxic peptides. Among them only 15–20 are used for feeding and the rest is mainly used for defense (http://penelope.uchicago.edu/~grout/encyclopedia_romana/aconite/geographus.html);

μ-Conotoxin GIIIA of the venom of this species has been reported to block with very high selectivity the muscle subtype of sodium channels;

Geographutoxin II (GTX II), a novel polypeptide toxin reduces sodium currents in rat myoballs (http://shop.bachem.com/h-2738–1.html);

A superfamily-G1c, G1d, G1a;

α-conotoxin GID (Grosso et al., 2014);

μ-Conotoxins (Terlau and Olivera, 2004);

α-conotoxin GIA and α-conotoxin GII (http://www.ebi.ac.uk/interpro/signature/PS60014/proteins-matched;jsessionid=72887C4196F6CFF05E E34D33FCB8F5C0);

ω-conotoxin GVIA.

Conus textile **(Linnaeus, 1758)**

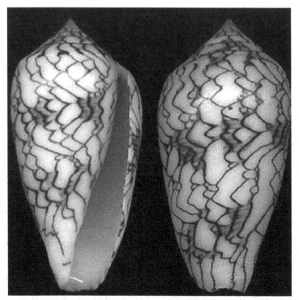

Apertural view   Abapertural view

Common Name: Textile cone, cloth of gold cone

Geographical Distribution: Red Sea, Indo-Pacific, Australia, New Zealand, the Indian Ocean from eastern Africa to Hawaii, and French Polynesia

Habitat: Intertidal range to depths of up to 50 m; coral reefs from the reef crest to deep waters of coral lagoons; in sand without vegetation; coral rubble, rock slabs, and dead coral in muddy substrate

Identifying Features: This species has a cone-shaped shell, which is usually cream patterned with goldish blotches (with darker brown cross-stripes) interspersed with irregular (mainly triangular) reddish-brown marks. This species grows between 40–150 mm. Tube-like siphon is whitish with reddish ring around tip and wide (5+ mm wide) blackish band about 1 cm back from tip. Proboscis is cream to pinkish. This species preys on many species of gastropods, small fishes, worms, dead cephalopods and shrimps.

Toxin Type:

Conotoxin of superfamilies, Contryphan, conopressin, T, A, M, O, I

Conotoxin α-4/6-conotoxin (CTx) TxID;

T-superfamily conotoxin TxVC;

TxIA, TxIB, TxI, TxIIA;

A-superfamily, conotoxin Tx1.1;

Conotoxin Tex31;

Conotoxin tx5a (Dovell, 2010);

Conotoxin tx9a (Aguilar et al., 2009);

Conootxin δ-TxVIA (Terlau and Olivera, 2004);

O-superfamily of conotoxins, Gla(1)-TxVI, Gla(2)-TxVI/A, Gla(2)-TxVI/B and Gla(3)-TxVI (Czerwiec et al., 2006);

Conotoxin Gla-TxXI, (http://www.ebi.ac.uk/interpro/entry/IPR020242/proteins-matched?start=40);

Contryphan-R/T, Leu-Contryphan-Tx and Contryphan-Tx (Gowd et al., 2005);

Arachidonic acid (Terlau and Olivera, 2004);

γ-glutamyl carboxylase (Terlau and Olivera, 2004);

Conus marmoreus (Linnaeus, 1758)

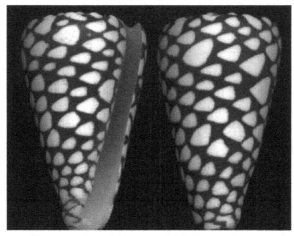

Apertural view   Abapertural view

Common Name: Marbled cone

Geographical Distribution: Indian Ocean along Chagos and Madagascar, in the Bay of Bengal along India; in the western part of the Pacific Ocean to Fiji and the Marshall Islands.

Habitat: In fairly shallow water (1–50 meters), typically on coral reef platforms or lagoon pinnacles, as well as in sand, under rocks or sea grass.

Identifying Features: Shell of this species is medium sized and moderately solid to heavy. Body whorl is conical and sides straight. About 10 to 12 weak and closely spaced spiral ridges are seen above the base. Body whorl is with weak regularly spaced spiral ribs on basal fourth to half. Shoulder is angulate and strongly tuberculate. Spire is low and its outline is straight. Ground color is white. Body whorl is with a regular network of black lines and triangular to rhomboid areas. Outlining is white tents that are quite uniform in shape and arrangement and usually separate from each other. In live specimen apex is purplish red. Aperture is white. This is a species which is believed to feed mostly on marine molluscs including other cone snails. Size of an adult shell can vary between 30 mm and 150 mm. These snails are predatory and venomous. As they are capable of "stinging" humans, live ones should not be handled.

Toxin Type: μO-conotoxins MrVIA and MrVIB (McIntosh et al., 1995);
  Conotoxin Mr038;
  I1-superfamily, conotoxin M11.2;
  Conotoxin Mr3.4 (http://www.ebi.ac.uk/interpro/signature/PS60014/proteins-matched;jsessionid=72887C4196F6CFF05EE34D33FCB8F5C0);
  Chi-conotoxin CMrX and CMrVIA (Dovell, 2010; http://www.uniprot.org/uniprot/P58809);
  Conotoxin mr5a (Dovell, 2010);
  Superfamily, Glacontryphan-M (Gowd et al., 2005);
  Conotoxins Mr1.1c, Mr1.8c and Mr001;
  T-superfamily conotoxins (Liu et al., 2012;)
  Some of the peptide toxins found in the venom of this species have been characterized and one of these is being developed as a potential drug for pain.

Conus striatus (Linnaeus, 1758)

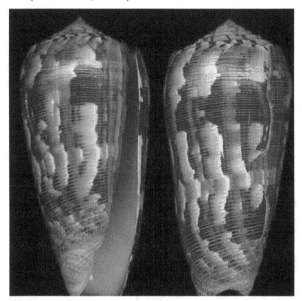

Apertural view   Abapertural view

Common Name: Hawaiian striated cone

Geographical Distribution: Red Sea, in the Indian Ocean along Aldabra Atoll, Madagascar, Mascarene Basin, Mauritius and Tanzania; in the Pacific Ocean along Philippines, Australia, New Zealand, New Caledonia and Thailand.

Habitat: Sand on coral reef, beneath rocks and dead coral slabs; at depths of 1–25 m.

Identifying Features: Shell of this species is irregularly clouded with pink-white and chestnut or chocolate, with fine close revolving striate, forming the darker ground-color by close colored lines. The pointed spire is tessellated with chestnut or chocolate and white. Its shoulders are rounded and its sutures are deep. Whorls are slightly channeled, carinate and striate. Outer lip is characterized by a posterior flare. Length of the shell may vary from 55 to 129 mm. This species preys on fish. These snails are predatory and venomous. As they are capable of "stinging" humans, live ones should be handled.

Toxin Type: κA-conotoxin family – κA-SIVA (Chen, 2008; Terlau and Olivera, 2004);

α-conotoxins SI, SIA, and SII (Terlau and Olivera, 2004);

A-superfamily, conotoxins S1, S1a;

δ-conotoxin SVIE (Montal, 2005);

Conotoxin S11.3 (http://www.ebi.ac.uk/interpro/entry/IPR020242/proteins-matched?start=40)

Conkunitzin-S1 (Dy et al., 2006).

Conus tulipa **(Linnaeus, 1758)**

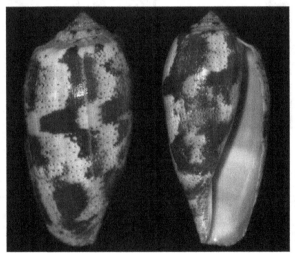

Abapertural view   Apertural view

Common Name: Tulip cone

Geographical Distribution: From Mozambique to Somalia across to the Marshall Islands and French Polynesia

Habitat: Intertidal areas of up to 10 m shallow waters; sand under coral reefs and among seaweed and rocky flats, which have been exposed to wave action, especially around the Great Barrier reef.

Identifying Features: Shell of this species is moderately large to large and moderately solid. Last whorl is ovate to cylindrical and outline is convex or almost straight and parallel-sided centrally. Left side is straight to distinctly concave at basal third. Aperture is wider at base than near shoulder. Shoulder is subangulate. Spire is low and it s outline is slightly concave or straight. Larval shell is of about 4 whorls with a maximum diameter 0.8–0.9 mm. First 4–7 postnuclear whorls are tuberculate. Teleoconch sutural ramps are usually somewhat concave, with 1 increasing to 4–9 spiral grooves. Latest whorls are usually with additional spiral striae and first 3 whorls are with a prominent subsutural ridge. Last whorl is with a few weak, variably spaced spiral ribsat base. Ground color is bluish gray and is suffused with blue or pink. Last whorl is with confluent reddish brown flames and blotches often fusing into an interrupted spiral band on each side of center. Spiral rows are of alternating brown and white dots and dashes from base to shoulder. Larval whorls are red, with a brown sutural line. Shell size varies from 50 to 95 mm. It feeds on fish and molluscs. Like all species within the genus Conus, these snails are predatory and venomous. They are capable of "stinging" humans and therefore live ones should be handled carefully or not at all.

Toxin Type: Conotoxin rho-TIA;
    Conantokin-T;
    Conotoxin TVIIA (Hill et al., 2000).

Conus ermineus (Born, 1778) (= *Chelyconus ermineus*)

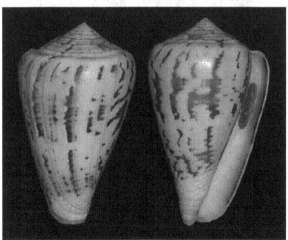

Abapertural view   Apertural view

Common Name: Turtle/Agate cone

Geographical Distribution: Western and eastern sides of the Atlantic Ocean; Cape Verdes; NW Africa; Mexico – Surinam; St. Vincent

Habitat: At depths of 0–101 m

Identifying Features: Shell of this species is normally covered by a thin, translucent, and light brown periastracum. In more mature specimens, this periastracum may be opaque, with widely spaced tufted spiral ridges, which become quite dense at the anterior end. It is a fishing eating species and it uses its specialized hollow harpoon like radula tooth to harpoon small fish and paralyze them with venom to facilitate swallowing. Shell size varies from 24 to 103 mm. These snails are predatory and venomous. As they are capable of "stinging" humans, live ones should not be handled

Toxin Type: Alpha-conotoxin EI; E.1.3; Conantokin-E; δ-conotoxin EVIA (δ-EVIA), a conopeptide in Conus ermineus venom that contains 32 amino acid residues and a six-cysteine/four-loop framework (Barbiera et al., 2004). This species along with *C.regius* and *C.centurio* has been to be potentially dangerous in Brazil (Junior et al., 2006).

Conus aulicus **(Linnaeus, 1758)**

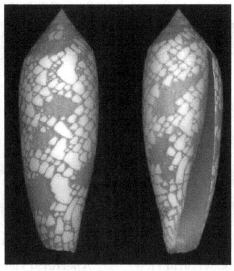

Abapertural view   Apertural view

Common Name: Princely cone,

Geographical Distribution: throughout the Indo-Pacific from East Africa to French Polynesia and the Marshall Islands

Habitat: Exposed coasts with heavily aerated water

Identifying Features: Shell of this species is moderately solid to heavy. Body whorl is broadly fusiform and its outline is straight. Shoulder is rounded. Spire is of moderate height and its outline is straight. Body whorl is with fine, closely spaced spiral ribs on basal fourth to third. Aperture is wider at base and outer lip is thick. Ground color is white. Body whorl is with irregular reddish brown and tan blotches separated by large white tents that tend to form three spiral bands, one below the shoulder, one at center and one above the base. Aperture is yellowish pink and outer lip is creamy yellow. Shell size varies from 65 to 163 mm. It preys upon gastropods. Like all species within the genus Conus, these snails are predatory and venomous. They are capable of "stinging" humans and therefore live ones should not be handled.

Toxin Type: au5a (Dovell, 2010); Alpha-conotoxin AuIB; AuIB (Adams and Berecki, 2013); Au1a, Au1b and Au1c.

Conus catus **(Hwass in Bruguière, 1792)**

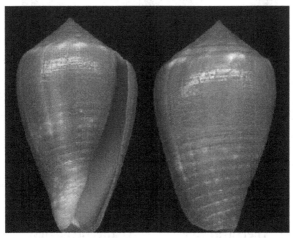

Apertural view   **Abapertural view**

Common Name: Cat cone

Geographical Distribution: Red Sea, Indian Ocean along the Aldabra Atoll, Chagos, Madagascar, Mauritius, Tanzania and KwaZulu-Natal; Indo-West Pacific Region

Habitat: Intertidally to depths of 8 m

Identifying Features: Shell of this species is bulbous, with a convex, striate spire. Body whorl is striate and the striae are rounded. It is usually obsolete above, granular below, olive, chestnut-, chocolate- or pink-brown, variously marbled and flecked with white, often faintly white-banded below the middle. It feeds mainly on fishes. Size of an adult shell varies between 24 mm and 52 mm. Like all species within the genus Conus, these snails are predatory and venomous.

Toxin Type: Omega-conotoxin CVIA; ω-conotoxin CVID (Terlau and Olivera, 2004); ω-conotoxins (CVIA–D) (Lewis et al., 2000).

Conus gloriamaris (**Chemnitz, 1777**) (= *Cylindrus gloriamaris*)

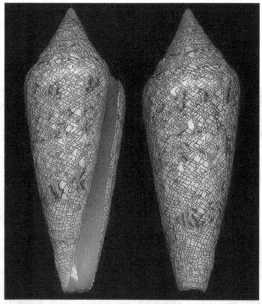

Apertural view   Abapertural view

Common Name: Glory of the sea cone, Glory of' gang of cones, spired cone

Geographical Distribution: Indo-Pacific; Philippines, Indonesia, Papua New Guinea, Sabah Malaysia, around the Solomon Islands, Vanuatu, Samoa and Fiji; Mozambique

Habitat: Sand and mud between 5 m and 300 m deep

Identifying Features: This species has a relatively large, slender shell with a tall spire. It is finely reticulated with orange-brown lines, enclosing triangular spaces similar to other textile cones, and two or three bands of chestnut hieroglyphic markings across its body. Its tan coloration can vary from a lighter, golden color to a deeper dark brown. Shell size varies from 75 to 160 mm. It feeds on molluscs. Once mature, it can reach a size ranging from 75 mm to 174 mm. It is inferred from protoconch morphology that this species has planktotrophic larval development. These snails are predatory and venomous. As they are capable of "stinging" humans, live ones should not be handled.

Toxin Type: β-superfamily – Conantokins; δ-Conotoxin GmVIA (Shon et al., 1994); Conotoxin gm9a (Aguilar et al., 2009); Conotoxin Gm9.1

Conus imperialis **(Linnaeus, 1758)**

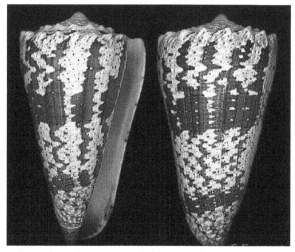

Apertural view   **Abapertural view**

Common Name: Imperial cone

Geographical Distribution: Entire Indo-Pacific except for Red Sea

Habitat: From intertidal to 240 m on sand, dead coral, coral rubble, reef limestone with or without algal turf, coral reef platform and in lagoons

Identifying Features: Shell of this species is moderately large to large and solid to heavy. Last whorl is conical and outline is largely straight, variably convex adapically. Shoulder is angulate strongly to sometimes weakly tuberculate. Spire is usually low and outline is slightly concave to slightly sigmoid. It is often with domed early postnuclear whorls and a projecting larval shell is surmounting an otherwise flat spire. Postnuclear spire whorls are distinctly tuberculate. Teleoconch sutural ramps are also variably concave. There are 4–10 spiral striae on late ramps. Last whorl is with weak to obsolete spiral ribs at base. Color of the shell is yellowish white, with numerous interrupted revolving lines and spots of dark brown and two irregular light brown bands. Aperture is white to violet, except for a dark violet to brown base, rarely extending to shoulder along outer margin. Size of the shell varies from 50 to 110 mm. It feeds exclusively on polychaetes. These snails are predatory and venomous. As they are capable of "stinging" humans, live ones should not be handled.

Toxin Type: α-Conotoxins, ImI and ImII (Grosso et al., 2014; Terlau and Olivera, 2004);
    K-superfamily, Conotoxin im23a and im23b (Ye et al., 2012);
    T-superfamily conotoxins (Liu et al., 2012);
    Kappa-conotoxin-likeIm11.3,ConotoxinIm11.4,Im11.2andIm11.1(http://www.ebi.ac.uk/interpro/entry/IPR020242/proteins-matched?start=40);
    Serotonin from the venom of C. imperialis (Terlau and Olivera, 2004);

Conus regius (Gmelin, 1791) (= *Stephanoconus regius*)

Apertural view   Abapertural view

Common Name: Crown cone, Royal cone

Geographical Distribution: Across the Caribbean and southern Gulf of Mexico from Atlantic coast of Florida, Florida Keys, Dry Tortugas; Mexico (Quintana Roo), Honduras, Costa Rica, Panama, Colombia, Venezuela including Islas Los Roques and islands of Cuba, Jamaica, Puerto Rico, Virgin Islands and Barbados then south to Brazil.

Habitat: Rocky, bouldery benthic sediment and coral reefs at depths to 20 m

Identifying Features: Adults are growing to 75 mm. Like all species within the genus Conus, these snails are predatory and venomous. They are capable of "stinging" humans and therefore live ones should be handled carefully or not at all. Nothing much is known about its biology.

Toxin Type: A- superfamily, conotoxin α-RgIB;

α-superfamily-α-conotoxin RgIA (http://www.ebi.ac.uk/interpro/signature/PS60014/proteins-matched;jsessionid=72887C4196F6CFF05EE3 4D33FCB8F5C0);

Conotoxin Rg91 (Aguilar et al., 2009);

Conotoxin reg3b (Dovell, 2010);

an α4/7-conotoxin RegIIA (Franco et al., 2012);

Conotoxin Reg12a (http://www.cusabio.com/Recombinant-Protein/Recombinant-Conus-regius–Conotoxin-Reg12a-193339.html);

Alpha-conotoxin RgIA (http://www.uniprot.org/uniprot/P0C1D);

This species along with *C.ermineus* and *C.centurio* s has been reported to be potentially dangerous in Brazil (Junior et al., 2006).

Conus litteratus (Linnaeus, 1758)

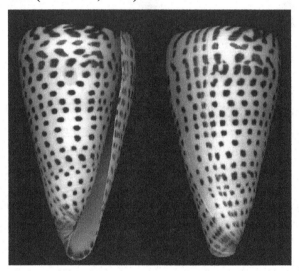

Apertural view   Abapertural view

Common Name: Lettered cone

Geographical Distribution: Indo-Pacific, except for Red Sea and Hawaiian Archipelago.

Habitat: Subtidal depths to 50 m; shallow waters; fine or coarse sand and rubble substrata with vegetation of varying density

Identifying Features: Shell of this species is moderately large to large and solid to heavy. Last whorl is conical and outline is almost straight, sometimes convex below shoulder. Base is moderately pointed. Shoulder is sharply angulate. Spire is low and its outline is usually sigmoid. Middle and late teleoconch sutural ramps are concave, with 3 increasing to 4–6 spiral grooves, which are often weak to obsolete. Last whorl is almost smooth. Ground color is white. Last whorl is usually encircled with 3 yellowish orange bands, at center and within adapical and abapical thirds. Spiral rows are of blackish brown and medium-sized, round to squarish spots are seen. Shell size varies from 60 to 170 mm. These snails are predatory and venomous. As they are capable of "stinging" humans, live ones should not be handled.

Toxin Type: α-superfamily, conotoxin Lt1.1 and Lt1A;

Conotoxin lt14a (with a unique cysteine pattern) (Peng et al., 2006);

Conotoxin lt9a-*C. litteratus* (Aguilar et al., 2009);

M- superfamily, conotoxin lt3a;

M-superfamily, conotoxin lt16.1 (Elliger et al., 2011);

L- superfamily, conotoxin lt14a (Elliger et al., 2011);

Conotoxin lt5d (Dovell, 2010);

Conotoxins lt11.2, lt11.3 and lt11.6 (http://www.ebi.ac.uk/interpro/entry/IPR020242/proteins-matched?start=40);

Conus lividus **(Hwass in Bruguière, 1792)**

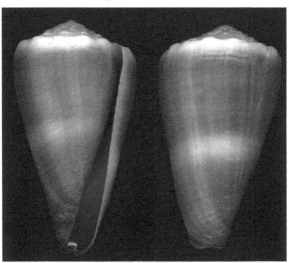

Apertural view   Abapertural view

Common Name: Livid cone; Olive- green cone

Geographical Distribution: Red Sea; in the Indian Ocean along Aldabra, Chagos, Mascarene Basin, Mauritius, Mozambique, Tanzania and the West Coast of South Africa; in the entire Pacific Ocean

Habitat: Intertidal, although more commonly found in subtidal regions with coral reefs up to a depth of 25 m; in sandy and rocky areas or in coral rubble, algal turf, reef limestone and bare reef limestone

Identifying Features: Shell of this species is small to moderately large and solid to heavy. Body whorl is broadly conical and outline is almost straight. Shoulder is angulate and is strongly to weakly tuberculate. Spire is of low to moderate height and its outline is straight to slightly concave. Body whorl is with variably granulose spiral ribs above base, sometimes to center. Body whorl is olive to brownish yellow, with pale or white transverse bands at center and below shoulder. Base is dark violet-brown. Apex is usually pink. Late spire whorls and shoulder are white, sometimes with paler ground color of body whorl between tubercles. Aperture is deep purple-violet. It is a vermiverous species that feeds on polychaetes and hemichordates. Adults are generally between 30 and 81 mm. These snails are predatory and venomous. As they are capable of "stinging" humans, live ones should not be handled

Toxin Type: Alpha-conotoxin (4/7 conotoxin), Lv1A (Elliger et al., 2011);

Conotoxin I2 (http://www.ebi.ac.uk/interpro/entry/IPR020242/proteins-matched?start=40);

Conotoxin Lv15a (http://www.uniprot.org/uniprot/C8CK76).

Conus magus **(Linnaeus, 1758)**

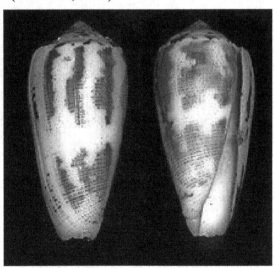

      Abapertural view   **Apertural view**

Common Name: Magical cone; magician's cone snail, magus cone

Geographical Distribution: Red Sea; in the Indian Ocean along Madagascar and the Mascarene Basin; in Pacific Ocean from Indonesia to Japan and to the Marshall Islands, Wallis and Futuna and Fiji, but mainly centered on the Philippines.

Habitat: Intertidal and upper subtidal depths of up to 100 m (for juveniles) in sandy coral reef and shallow bays. Prefers to hide beneath rocks and dead corals.

Identifying Features: This is a common species with very variable in pattern and shade of coloring Spire is moderate and striate. Body whorl is long and rather cylindrical, closely striate below. Color of the shell is white, clouded with bluish ash, orange-brown, chestnut or chocolate, everywhere encircled by narrow chocolate interrupted lines, often separated into somewhat distant dots Middle of the body whorl is usually irregularly fasciate with white. Spire is tessellated with chestnut or chocolate. Adults prey on fish and are active at night, whereas juveniles are vermivorous. Size of an adult shell varies between 16 mm and 94 mm. These snails are predatory and venomous. As they are capable of "stinging" humans, live ones should not be handled

Toxin Type: ω-conotoxin MVIIA-Conus magus (Terlau and Olivera, 2004);
   ω-conotoxin MVIIC, Conus magus (Terlau and Olivera, 2004);
   Conodipine, a novel phospholipase A2 (Terlau and Olivera, 2004);
   Conodipine-M (Elliger et al., 2011);
   α-superfamily, conotoxin M1;
   α-conotoxin MI (http://www.ebi.ac.uk/interpro/signature/PS60014/proteins-matched;jsessionid=72887C4196F6CFF05EE34D33FCB8F5C0);
   ω-conotoxin FVIA – It is a highly reversible conotoxin targeting N-type Ca^{2+} channels with analgesic effect. Though CTx-MVIIA (Ziconotide) of this species has been reported to depress arterial blood pressure immediately after administration, this pressure recovers faster and to a greater degree after CTx-FVIA administration.

   Ziconotide (a synthetic version of ω-conotoxin M VII A), an analgesic drug derived from the toxin of this species acts as a painkiller with a potency 1000 times that of morphine. Ziconotide works by blocking sodium channels in pain-transmitting nerve cells, rendering them unable to transmit pain signals to the brain. It is administered through injection into the spinal fluid. It has been developed for treatment of chronic and intractable pain caused by AIDS, cancer, neurological disorders and other maladies, and approved by the U.S. Food and Drug Administration in December 2004 under the name Prialt (Sheeja et al., 2011). Though Ziconotide alleviates pain effectively (without causing addiction, by blocking the pores of sodium channels). Unfortunately, CTx-MVIIA has a narrow therapeutic window and produces serious side effects due to the poor reversibility of its binding to the channel. It would thus be desirable to identify new analgesic blockers with binding characteristics that lead to fewer adverse side effects.

Conus obscurus (G. B. Sowerby I, 1833)

Apertural view   Abapertural view

Common Name: Dusky cone, obscure cone

Geographical Distribution: Indian Ocean along Aldabra, the Mascarene Basin and Tanzania; in the Pacific Ocean to Hawaii and French Polynesia.

Habitat: Intertidal to more than 40 m usually around coral reefs; various reef substrata (caves among coral heads, coral rubble among sea-weed, and on patches of sand, as well as on intertidal rocky flats)

Identifying Features: Shell of this species is moderately small to medium-sized and light to moderately light. Last whorl is cylindrical to narrowly cylindrical. Outline almost straight. Left side is slightly concave at basal fourth to third. Aperture is wider at base than near shoulder. Shoulder is angulate. Spire is usually of moderate height and its outline is straight or slightly concave. Larval shell is of 3.75–4.0 whorls. First 2–4 postnuclear whorls are tuberculate. Teleoconch sutural ramps are flat to slightly concave, with 1 increasing to 4–6 major spiral grooves, sometimes to 8 finer grooves. Sculpture is usually weak on late ramps. Last whorl is with a few weak spiral ribs at base. Last whorl is grayish blue to violet. Variably prominent spiral rows of alternating brown and gray dots and dashes are seen from base to shoulder. Brown flames, clouds and blotches are usually concentrated in spiral bands below shoulder, just above center and within basal third. Larval whorls are red to orange. Early postnuclear sutural ramps are gray and are often dotted with dark brown at both margins in first whorl and with brown radial lines and streaks in following whorls. Later sutural ramps are bluish gray with confluent brown radial blotches, often completely overlaid with brown. Aperture is translucent or with thin white enamel. Shell size varies from 25 to 44 mm. Like all species within the genus Conus, these snails are predatory and venomous. They are capable of "stinging" humans and therefore live ones should be handled carefully or not at all.

Toxin Type: Alpha-conotoxin OIVA (http://www.uniprot.org/uniprot/P69746);

alphaA-conotoxin OIVB (alphaA-OIVB), (http://www.pubfacts.com/search/alphaA-conotoxin+OIVB+Conus+obscurus).

Conus omaria (Hwass in Bruguière, 1792)

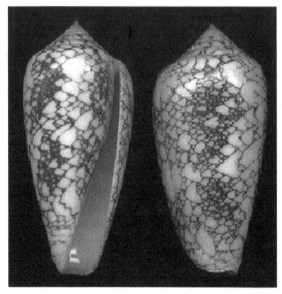

Apertural view   Abapertural view

Common Name: Omaria cone

Geographical Distribution: Indian Ocean along Aldabra, Madagascar, the Mascarene Basin and Tanzania

Habitat: Shallow subtidal habitats (10–100 m) on coral reefs and in reef lagoons, lagoon interisland reefs and reef pinnacles; sand and rubble, under rocks

Identifying Features: Color of the shell of this species arise from orange-brown to chocolate-color, covered by minute white spots, and overlaid by larger white triangular spots, sometimes forming bands at the shoulder, middle and base. It feeds on other gastropods during night time only. Shell attains a length of 60 mm. Like all species within the genus Conus, these snails are predatory and venomous. They are capable of "stinging" humans and therefore live ones should be handled carefully or not at all.

Toxin Type: A-superfamily, Om1a

Conus pennaceus (Born, 1778)

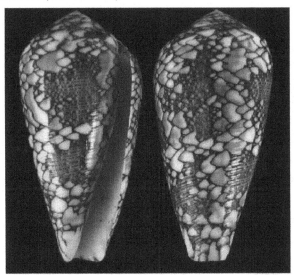

Apertural view   Abapertural view

Common Name: Feathered cone

Geographical Distribution: Red Sea, in the Indian Ocean along Madagascar, the Mascarene Basin and Tanzania

Habitat: Shallow water coral reefs, on rubble, in sandy substrate, and in sandy/ muddy substrate

Identifying Features: Color of the shell of this species varies from orange-brown to chocolate, covered by minute white spots, and overlaid by larger white triangular spots, sometimes forming bands at the shoulder, middle and base. Size of an adult shell varies between 35 mm and 88 mm. Like all species within the genus Conus, these snails are predatory and venomous. They are capable of "stinging" humans and therefore live ones should be handled carefully or not at all.

Toxin Type: alpha-conotoxin PNI1;
(http://www.conoserver.org/?page=card&table=structure&id=12);
A-superfamily, α-Conotoxins PnIA and PnIB (Terlau and Olivera 2004);
γ-conotoxin PnVIIA) and μ-PnIVA (Terlau and Olivera, 2004);

T-superfamily, conotoxins Pn5.1, Pn10.1, Pn5.2, Pn-0111, Pn-014, PnMRCL-012;

M-superfamily, conotoxins Pn3.2, Pn3.1;

O1-superfamily, conotoxins Pn6.6, Pn6.7 and Pn6.12, Pn6.13, Pn6.14, Pn6.5, Pn6.3, Pn6b, Pn6a;

A-superfamily, conotoxins PnMGMR-02, PnIB;

O2-superfamily, conotoxins Pn6.8, Pn6.9, Pn7a;

Conotoxins Pn3.4, Pn3.3, Pn4.1, Pn4a;

O3-superfamily, conotoxins Pn6.10, Pn6.11; (http://www.conoserver. org/index.php?page=list&table=protein&Organism_search%5B%5D= Conus+pennaceus&Type%5B%5D=Wild+type&sort_by=Class).

Conus pulicarius (Hwass in Bruguière, 1792)

Apertural view　　　　**Abapertural view**

Common Name: Flea cone

Geographical Distribution: Central and Western Pacific including Papau New Guinea, the Philippines, and China. Also French Polynesia, Hawaii, the Indian Ocean including Cocos (Keeling) island, and North Western Australia.

Habitat: Intertidal range to around 75 m

Identifying Features: Shell of this species is medium-sized to moderately large and solid to heavy. Last whorl is conical, conoid-cylindrical, or ventricosely conical. Outline is convex at subshoulder area and almost straight below, often with slight convexity above base. Shoulder is subangulate to rounded, weakly to strongly tuberculate. Spire is of low to moderate height and outline is slightly concave to straight. Larval shell is of about 3.5 whorls. Postnuclear spire whorls are strongly tuberculate. Teleoconch sutural ramps are concave, with 1 increasing to 4–5 spiral grooves. Last whorl is with variably spaced spiral grooves and adjacent ribs on basal fourth. Ground color is white. Last whorl is with spiral rows of irregularly set black spots or bars clustered in an interrupted spiral band within adapical and abapical third. Clusters are often emphasized by underlying shadows of yellow, brown or violet. Near base, white dashes usually alternate with black markings. Black spots and bars may fuse into solid axial flames. Larval whorls are white to gray. Teleoconch sutural ramps are with variously solid black markings, varying in number and arrangement. Aperture is white to bluish white, often suffused with yellow or orange. Shell size varies from 35 to 75 mm. Like all species within the genus Conus, these snails are predatory and venomous. They are capable of "stinging" humans and therefore live ones should be handled carefully or not at all.

Toxin Type: Conotoxin sequences, belonging to at least 14 superfamilies viz, A, M, T, O, S, P, I, J, L, V, conantokin, contryphan, contulakin and conkunitzin have been identified from this species (Lluismaa et al., http://content.lib.utah.edu/utils/getfile/collection/uspace/id/2546/filename/image);

α-conopeptides in the salivary gland.(Kendel et al., 2013);

Synthetic conotoxin pu14a (induces a sleeping phenotype in mice, and it is also toxic to freshwater goldfish upon intramuscular injection);

m-conotoxin;

Pu3.1 (http://www.conoserver.org/?page=list&table=protein&Organism_search%5B%5D=Conus%20pulicarius&Type%5B%5D=Wild+type – Cono Server);

Conantokin-Pu1 & Pu2;

Conkunitzin-Pu1 & Pu2;

Contryphan-like Pu1, Pu2 and Pu3;

Contryphan-Pu1 and Pu2;

Contulakin-Pu;

Conotoxins Pu1.1 to Pu1.6, Pu11.1, Pu11.10, Pu11.12, Pu11.2 to Pu11.9, Pu11.2, Pu1, Pu11.4, Pu11.5, Pu11.6 to Pu11.8, Pu11.9, Pu14.2 to Pu14.7, Pu14.2, Pu14.3, Pu14.4 to 14.6, Pu14.7, Pu14a, Pu15.1, Pu19.1, Pu3.1 to Pu3.7, Pu6.1, Pu6.2, Pu6.3, Pu6.4 to Pu6.6, Pu6.7, Pu6.8 & Pu 6.9, Pu6.10 & 6.11, Pu12 & Pu13, Pu14, Pu15 & Pu16, Pu17, Pu18 to Pu37, P5.1 & Pu5.10, Pu5.12 to Pu5.19, Pu9.1 & Pu9.2, Pu9.3, PuIA & PuIIA, PuSG1.1 & Pu1.2.

Conus quercinus (Lightfoot, 1786)

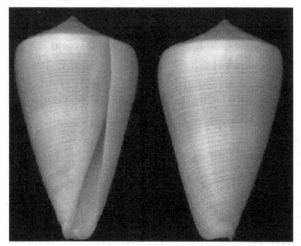

Apertural view   **Abapertural view**

Common Name: Oak cone

Geographical Distribution: Throughout the entire Indo-Pacific

Habitat: Sand-dwelling species, living on sand flats within bays and avoids rocky, coral and limestone outcrops; among vegetation.

Identifying Features: Shell of this species is moderately large, heavy and low gloss. Body whorl is broadly conical and its outline is slightly convex adapically and straight below. Body whorl is with few low spiral ridges above base, continuing as spiral threads above center. Shoulder is broad and subangulate. Spire is of moderate height and its outline is slightly concave. Aperture is wide, slightly flaring at base and outer lip is straight and thick. Ground color is pale yellow. A paler mid-body band is visible. Spire is uniformly yellowish white and early whorls are dark brown. Aperture is white. Adults of this species are typically between 60–140 mm in length. Like all species within the genus Conus, these snails are predatory and venomous. They are capable of "stinging" humans and therefore live ones should be handled carefully or not at all.

Toxin Type: Conotoxins Cq51–1, Cq51–2, Cq53–1, Cq53–2, CQ55–1, CQ55–2, CQ39 (Chen, 2008);

α-conotoxin Qc1.2 (Grosso et al., 2014);

α4/4-conotoxin, Qc1.2;

m-conotoxin Qc3.1.

Conasprella centurio **(Born, 1778)** *(= Conus century, Kohniconus centurio)*

Apertural view   **Abapertural view**

Common Name: Centurion cone

Geographical Distribution: Throughout the Caribbean from the Bahamas to southern Brazil

Habitat: Gravel and sand, and also in coral rubble and on coral reefs from shallow water to 100 m.

Identifying Features: Shell of this species is medium sized to large, moderately solid or solid to moderately heavy. Last whorl is conical to broadly conical. Outline is straight to slightly concave. Shoulder is carinate. Sire is moderate to high, straight or slightly concave in outline. Teleconch sutural ramps are concave. Last whorl is with low 4–7 rounded spiral ribs basally and 6–10 broader rounded ribs separated by finely axially threaded grooves adapically. Shell size varies from 30 to 100 mm. Like all species within the genus Conasprella, these cone snails are predatory and venomous. They are capable of "stinging" humans and therefore live ones should be handled carefully or not at all.

Toxin Type: Conopeptides; venom type not reported. In Brazil, this species along with *C.regius* and *C.ermineus* have been reported to be potentially dangerous (Junior et al., 2006).

3.1.2   *VENOMOUS CONE SNAILS*

***Conus abbreviatus* (Reeve, 1843) (=*Miliariconus abbreviatus*)**

Apertural view   Abapertural view

Common Name: Abbreviated cone

Geographical Distribution: Endemic to Hawaiian Islands

Habitat: Reefs and tide pools to 20 m

Identifying Features: Color of the shell is tan, olive or gray with small brown dots and coronated spire. Animal has a pink siphon and feeds exclusively on polychaetes, mainly Eunicidae and Nereidae. Shell attains a height of 5 cm. Like all species within the genus Conus, these snails are predatory and venomous. They are capable of "stinging" humans and therefore live ones should not be handled.

Toxin Type: Conotoxin AbVIL-four-loop conotoxin (Duda and Remigio, 2008) and recombinant Conus abbreviatus conotoxin AbVID.

Conus achatinus **(Gmelin, 1791)** **(=** *Pionoconus achatinus, Conus monachus***)**

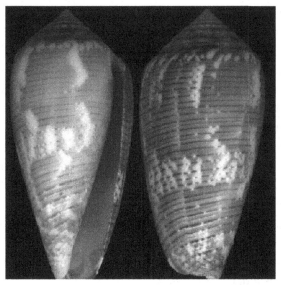

Apertural view   Abapertural view

Common Name: Turtle cone

Geographical Distribution: Indo-Pacific region, ranging from Mozambique and Tanzania in the west to the Philippines, Solomon Islands and New Caledonia in the east.

Habitat: Intertidal and subtidal areas up to a depth of 20 m. It is found in sand and mud under rocks, coral rubble and under coral

Identifying Features: Shell is bulbous, with a somewhat elevated, lightly striated spire and rounded shoulders. Body whorl is rounded with convex sides, sometimes with granular striae below. Shell is pale blue, marbled with pinkish or purplish white and olivaceous-brown. There are also close-set narrow brown lines, which are usually broken up into brown and white articulations. Aperture is bluish white. End of siphon is dark brown and proboscis is deep orange to red. Adults can grow to a maximum length of 100 mm. It has the largest eggs of any known Indo-Pacific species. These snails are predatory and venomous. As they are capable of "stinging" humans live ones should not be handled.

Toxin Type:

    (i) Alpha-conotoxin-like Mn1.4 (http://www.ebi.ac.uk/interpro/signature/PS60014/proteins-matched;jsessionid=72887C4196F6CFF05EE34D33FCB8F5C0);

    (ii) Pentadecamer peptides, Ac1.1a and Ac1.1b, with appropriate disulfide bonding; These peptides provide additional tools for the study of the structure and function of nicotinic receptor (Liu et al., 2007);

    (iii) Kappa-conotoxin Ac4.2;

    (iv) Delta-conotoxin-like Ac6.1. Alpha-conotoxin-like Ac1.1b, Omega-conotoxin-like Ac6.4;

    (v) Alpha-conotoxin-like Ac1.1b (http://www.ebi.ac.uk/interpro/signature/PS60014/proteins-matched;jsessionid=72887C4196F6CFF05EE34D33FCB8F5C0);

    (vi) *C.achatinus* Ac6.1, Ac6.2, Ac6.3 and Ac6.4 (Ramasamy and Manikandan, 2011);

    (vii) Conus achatinus alpha conotoxin Ac4.3b;

    (viii) Conus achatinus conotoxin Ac1.1a.

Conus adamsonii (**Broderip, 1836**) (= *Conus aureolus, Conus casta-neus, Conus cingulatus, Conus rhododendron*)

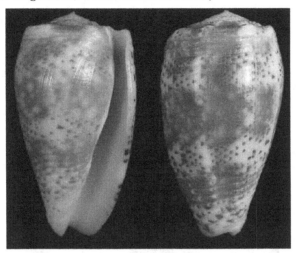

Apertural view   Abapertural view

Common Name: Rhododendron Cone

Geographical Distribution: S. Pacific from Coral Sea to French Polynesia

Habitat: Intertidal; 10 to 50 m; on seeward sides of coral reefs and in lagoons; on large stretches or small pockets of sand

Identifying Features: This species has a medium-sized, moderately solid-to-solid shell. Last whorl is ventricosely conical or conoid-cylindrical to ovate. Outline is variably convex adapically, less so (right side) or slightly concave (left side) toward base. Aperture is wider at base than near shoulder. Shoulder is subangulate to angulate. Spire is low and its outline is straight to concave. Ground color is white and is partially suffused with pink to purple. Last whorl is with 3 rather broad spiral bands of confluent violet or brown nebulous flecks and tent-like spots, below shoulder, just above center, and within abapical third, alternating with 3–4 rather narrow spiral zones of very small brown to dark reddish or purplish brown triangular spots. Color bands contain prominent to obsolete spiral rows of irregularly alternating white and brown dots and dashes. Adults can grow to 56 mm although they will typically be about 35 mm. Like all species within the genus Conus, these snails are predatory and venomous. They are capable of "stinging" humans and therefore live ones should not be handled.

Toxin Type: Conohyal-Cn1, (Hyal sequence) has been reported to occur in the venom duct transcriptome of this species (Violette et al., 2012).

Leptoconus amadis **(Hwass in Bruguière, 1792)** (= *Conus amadis*)

Abapertural view   Apertural view

Common Name: Amadis cone

Geographical Distribution: Mascarene Basin, in the Indian Ocean and in the Pacific Ocean along Indonesia, New Caledonia and Polynesia.

Habitat: Intertidal mudflats to a depth of 18 m in sand.

Identifying Features: Spire of this species is striate, channeled, concavely elevated and sharp-pointed. It has a sharp shoulder angle. Lower part of body whorl is punctured and grooved. Color of the shell is orange-brown to chocolate and is thickly covered with large and small subtriangular white spots, which by their varied disposition sometimes form a white central band, or dark bands above and below the center, the latter occasionally bearing articulated revolving lines. This species grows to a maximum of 110 mm. Like all species within the genus Conus, these snails are predatory and venomous. They are capable of "stinging" humans and therefore live ones should not be handled. Toxins isolated from this species show hemolytic activity.

Toxin Type: Am2766 (which has been reported to target mammalian sodium channel) and Contryphan-P/Am (Gowd et al., 2005); Am975 (Ramasamy and Manikandan, 2011; Sarma et al., 2005; Gowd et al., 2005); am9a (Aguilar et al., 2009).

Leptoconus ammiralis **(Linnaeus, 1758)** *(= Conus ammiralis)*

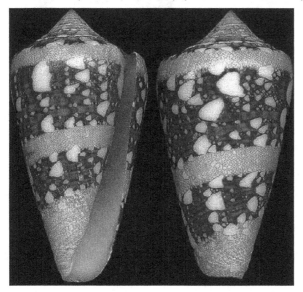

Apertural view   Abapertural view

Common Name: Admiral cone

Geographical Distribution: From Mozambique to Kenya and also in the Seychelles, Reunion, and Andaman Islands, east to the Marshall Islands and Fiji, including northern Australia, and north to Japan

Habitat: Fine, coarse sand, beneath rocks and among algae at depths of between 2 and 240 m

Identifying Features: The shell of this species has the characteristic golden stripes and it grows to a maximum size of 72 x 40 mm. Like all species within the genus Conus, these snails are predatory and venomous. They are capable of "stinging" humans, and therefore live ones should not be handled.

Toxin Type: Alpha-conotoxin-like Ai1.2 (http://www.uniprot.org/uniprot/P0CB08).

***Gradiconus anabathrum* (Crosse, 1865) (= *Conus floridanus floridensis*)**

Apertural view   Abapertural view

Common Name: Florida/Mrs. Burry's/Philippi's Cone

Geographical Distribution: Caribbean Sea and the Gulf of Mexico; Florida, USA; Cuba; Mexico

Habitat: 0–122 m; outer coral reefs

Identifying Features: Spire of this species is elevated, and gradate. Body whorl is grooved towards the base. The color of the shell is pale yellowish brown, with a central white band and scattered white maculations, obscurely encircled by lines of light chestnut spots. It is a worm hunting species and feeds mainly on polychaetes. Maximum-recorded shell length is 51 mm. Like all species within the genus Conus, these snails are predatory and venomous. They are capable of "stinging" humans and therefore live ones should not be handled.

Toxin Type: Conotoxin flf14a (http://www.uniprot.org/uniprot/P84705);
Conotoxin flf14c (http://www.uniprot.org/uniprot/P84707);
Conotoxin flf14b (http://www.uniprot.org/uniprot/P84706; Moller et al., 2005).

Conus anemone (Lamarck, J.B.P.A. de, 1810)

Apertural view   **Abapertural view**

Common Name: Anemone cone

Geographical Distribution: Endemic to Australia; Southern Queensland to Houtman Abrolhos, WA, including Tasmania and Lord Howe Is.

Habitat: Low intertidal and subtidal on the open coast, down to 55 m; under rocks and ledges on rocky shores and on sand among algae or eel-grass.

Identifying Features: This is a variable species and shells are light to medium in weight, relatively tall with body whorl uniformly convex, or convex posteriorly and slightly concave towards the base. Entire body whorl is sculptured with spiral ridges, often weaker midbody. Spire is flat to medium in height and spire whorls are with 5–8 spiral threads. Aperture is wider anteriorly and outer lip is convex. Background color is bluish-gray, with irregular brown patches, which may coalesce into axial or spiral bands. Aperture is bluish-white in deep interior and brownish-purple at outer edge. This species has a maximum length of 76 mm. Like all species within the genus Conus, these snails are predatory and venomous. They are capable of "stinging" humans and therefore live ones should not be handled.

Toxin Type: Alpha-conotoxin AnIB (http://www.uniprot.org/uniprot/P0C1 V7);

Anla and Anlb;

Alpha-conotoxin-like AnIC (http://www.uniprot.org/uniprot/P0C1 V8);

Alpha-conotoxin AnIB and Alpha-conotoxin-like AnIC (http://www.ebi.ac.uk/interpro/signature/PS60014/proteins-matched;jsessionid=72887C4196F6CFF05EE34D33FCB8F5C0).

Conus araneosus (Lightfoot, 1786)

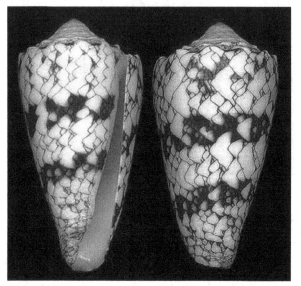

Apertural view   Abapertural view

Common Name: Cobweb cone

Geographical Distribution: In the Indian Ocean along India, Sri Lanka and Tanzania, and in the Pacific Ocean along the Philippines and Indonesia

Habitat: Intertidal to 20 m, on limestone and sandy substrata

Identifying Features: Shell of this species is moderately large and solid to heavy with a high gloss. Body whorl is conical. Shoulder is broad and faintly canaliculated, angulate and weakly to strongly tuberculate. Outline is straight to slightly convex. Spire is of low to moderate height and its outline is straight. Spiral whorls are strongly tuberculate. Body whorl is with weak spiral ribs above base. Aperture is moderately wide. Outer lip is thick, sharp and straight. Ground color is white and entire shell is tinged with violet. Body whorl is with fine network of dark brown reddish brown lines outlining small white tents, with two dark brown or black spiral bands on each side of the center. Aperture is white to pale violet and interior is deep yellow. Size of an adult shell varies between 48 mm and 100 mm. It appears to feed on other gastropods. Like all species within the genus Conus, these snails are predatory and venomous. They are capable of "stinging" humans and therefore live ones should not be handled.

Toxin Type: Chi-conotoxin-like Ar1311;

C.araneosus Ar1447, Ar1859 and Ar1813 (Ramasamy and Manikandan, 2011).

Puncticulis arenatus **(Hwass in Bruguière, 1792) (=** *Conus arenatus***)**

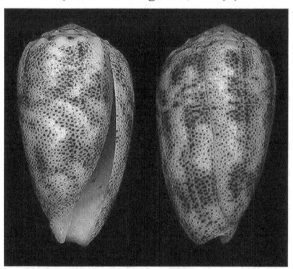

Apertural view   **Abapertural view**

Common Name: Sand-dusted cone

Geographical Distribution: Indo-West Pacific

Habitat: Reef flats

Identifying Features: Shell of this species is medium sized to large and moderately solid to moderately heavy. Body whorl is ventricosely Conical and outline is convex. Shoulder is rounded and indistinctly tuberculate.

Spire is low and outline moderately convex. Body whorl is with weak spiral ribs at base and ribs are granulose. Aperture is narrow posteriorly and wide anteriorly. Outer lip is thick and convex. Ground color is white. Body whorl is with spiral rows of widely spaced brown dots and clustered in two interrupted spiral bands. White dashes are often irregularly alternating with brown dots. Spire is with radial clusters of brown dots. Aperture is white. Shell size varies from 25 to 90 mm. Like all species within the genus Conus, these snails are predatory and venomous. They are capable of "stinging" humans and therefore live ones should not be handled.

Toxin Type: Conotoxin ArMLCL-D01 (http://www.uniprot.org/uniprot/Q9BH79)-ion channel inhibitor activity;

Alpha-conotoxin ArIA (http://www.uniprot.org/uniprot/P0C8R2)-showing acetylcholine receptor inhibitor activity;

Conotoxin ArMKLT2–0311(http://www.uniprot.org/uniprot/Q9BP83)—showing ion channel inhibitor activity.

Conus aristophanes **(G. B. Sowerby II, 1857)**

Apertural view   Abapertural view

Common Name: Cone snail

Geographical Distribution: Polynésie Française (Huahine)

Habitat: 50 m deep sandy and rocky islands

Identifying Features: This species is characterized by its only one or two coarse spiral ridges. Size of the shell is 35.2 × 23.2 mm. Like all species within the genus Conus, these snails are predatory and venomous. They are capable of "stinging" humans and therefore live ones should not be handled.

Toxin Type: Histone H3-Four-loop conotoxin (Duda and Remigio, 2008); Four-loop conotoxin (http://www.uniprot.org/uniprot/A9P3X5); Conotoxin- 01 superfamily At6.1 to At6.8 and At6.1 precursor to At6.8 precursors (http://www.conoserver.org/index.php?page=list&table=protein &Organism_search%5B%5D=Conus%20aristophanes).

Phasmoconus asiaticus **(Motta, A.J. da, 1985)** (= *Conus asiaticus*)

Apertural view   Abapertural view

Common Name: Not designated

Geographical Distribution: Pacific Ocean along the Philippines and Japan and in the South China Sea along Vietnam

Habitat: Not reported

Identifying Features: Shell of this species is medium sized and moderately solid. Body whorl is conical and outline is convex adapically. Left side is slightly concave near base. Shoulder is angulate and weakly granulose. Spire is of moderate height and its outline is concave. Body whorl is with strong, with prominent widely spaced spiral ribs and ribbons, and is strongly granulose. Ground color is white. Body whorl is overlaid with yellowish brown irregular axial streaks and blotches arranged in two spiral bands. Aperture is white. Size of an adult shell varies between 35 mm and 52 mm. Like all species within the genus Conus, these snails are predatory and venomous. They are capable of "stinging" humans and therefore live ones should not be handled.

Toxin Type: Conotoxin with framework III (CC-C-C-CC) and a molecular mass of 1702.04 Da and conotoxin with framework XIV (C-C-C-C) and a molecular mass of 1699.93 Da (Lebbe, et al., 2014).

Conus australis **(Holten, 1802)** (= *Graphiconus australis*)

Apertural view   Abapertural view

Common Name: Austral cone

Geographical Distribution: Japan to Philippines and Vietnam; India and W. Thailand; Fiji

Habitat: Mud and gravel; at depths of 35–240 m

Identifying Features: Shell of this species is moderately large to large and moderately solid-to-solid, with low gloss. Body whorl is narrowly conoid-cylindrical and its outline is convex. Shoulder is subangulate. Spire is of moderate height and is sharply pointed with slight concave outline. Body whorl is encircled with variably spaced granulose ribs, which are obsolete posteriorly. Aperture is moderately wide and outer lip is straight and thin. Ground color is white, suffused with pale brown. Body whorl is with brown blotches, tending to form three interrupted spiral bands, below the shoulder and above-and-below center, the former weakest. Most have many rows of widely spaced squarish brown dots on the spiral ribs. Aperture is pale violet. Shell size varies from 40 to 123 mm. Like all species within the genus Conus, these snails are predatory and venomous. They are capable of "stinging" humans and therefore live ones should not be handled.

Toxin Type: A 16-amino acid peptide namely α-conotoxin AusIA has been isolated from this species. This peptide has the typical α-conotoxin CC-Xm-C-Xn-C framework, but both loops (m/n) contain 5 amino acids, which has never been described before. α-conotoxin classification has been reported to be helpful in the design of novel therapeutic compounds. (Lebbe et al., 2014).

Conus bandanus (Hwass in Bruguière, 1792)

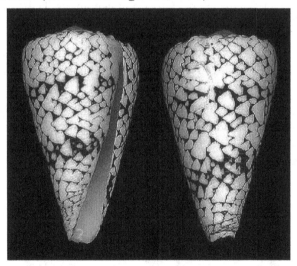

Apertural view   Abapertural view

Common Name: Banded marble cone

Geographical Distribution: Throughout the Indo-Pacific region, ranging from the East coast of Africa to French Polynesia and Hawaii, excluding the Arabian Sea and Bay of Bengal

Habitat: At depths of between 5 and 90 m on coral reef, reef lagoons, sand, among algae as well as under rocks and rubble.

Identifying Features: Shell of this species is moderately small to large and moderately light to heavy. Last whorl is conical to ventricosely conical and outline is nearly straight, variably convex adapically. Shoulder is angulate and moderately to strongly tuberculate. Spire is of low to moderate height. Teleoconch sutural ramps are concave in late whorls, with 2–4 weak spiral grooves and additional spiral striae. Spiral sculpture is often obsolete. Last whorl is with weak spiral grooves on basal third to three-fourths. Anterior end of aperture is violet-brown or orange-brown and rest of aperture is white, suffused with blue or orange. Ground color is white to pale violet or pale pink. Last whirl is with a blackish brown network of lines and triangular areas and rhomboid blotches are clustered in a spiral band on either side of the central area. Apex is white to light purple. Aperture is white, occasionally tinged with violet, pink or yellow and base of aperture may be brown. Shell size varies from 45 to 150 mm. Like all species within the genus Conus, these snails are predatory and venomous. They are capable of "stinging" humans and therefore live ones should not be handled.

Toxin Type: BnIIID conopeptide. This peptide belongs to the M-1 family of conotoxins. This is the first report of a member of the M-superfamily containing bromotryptophan (Nguyen et al., 2014); Alpha-conotoxin-like Bn1.3 (http://www.ebi.ac.uk/interpro/signature/PS60014/proteins-matched;jsessionid=72887C4196F6CFF05EE34D33FCB8F5C0).

Conus betulinus **(Linnaeus, 1758)**

Apertural view   Abapertural view

Common Name: Beach cone

Geographical Distribution: Widespread in the Indo-West Pacific, from East Africa to eastern Polynesia; north to southern Japan and south to Queensland and New Caledonia

Habitat: Sand flats, especially in sheltered areas and near sea grasses; littoral and shallow sublittoral zones to a depth of about 20 m.

Identifying Features: This species has a solid to heavy shell. Body whorl is broadly pyriform and outline is convex. It has about a dozen rounded and closely spaced basal spiral ridges and reminder of the whorl is smooth. Shoulder is very broad, rounded and not distinct from spire. Spire is very low, nearly flat except for the first few whorls, forming a small projecting pointed cone. Aperture is wide, nearly uniform and outer lip is thick, sharp and straight. Body whorl is uniformly creamy white, yellowish to deep orange and is covered with few to many spiral rows of small to large blackish spots, usually regularly spaced. Spots may be large squares or rounded dots, sometimes with secondary rows of finer brown dots between. Aperture is white and margins are pale orange. Maximum shell length is 17.5 cm. Like all species within the genus Conus, these snails are predatory and venomous. They are capable of "stinging" humans and therefore live ones should not be handled

Toxin Type: BeTXIa, BeTXIb, BeTXIIa, and BeTXIIb (Chen et al., 1999; Aguilar et al., 2009); Venom of this species shows hemolytic activity in erythrocytes Further it showed considerable enzymatic properties like gelatinolytic, caesinolytic, fibrinolytic and fibrinogenolytic activities besides exhibiting significant activity against HeLa cell lines; kappa-conotoxin (κ-BtX); bt5a.

Stephanoconus brunneus (Wood, W., 1828) (= *Conus brunneus*)

Apertural view　　Axial view　　Abapertural view

Common Name: Wood's brown cone

Geographical Distribution: Gulf of California, and Baja California, Mexico to Ecuador, Galapagos Islands, Revillagigedo Islands, Clipperton Island and Cocos Islands

Habitat: Rocks and rubble, and in sand channels at depths between 10 and 30 m

Identifying Features: Shell of this species is small, elongately conical, thin, and fragile. Shoulder is sharply angled and carinated. Spire is moderately elevated, with stepped whorls. Body whorl is shiny and polished and anterior tip is encircled with 6 small spiral cords. Aperture is narrow, and slightly wider at anterior end. Protoconch is proportionally large, and mamillated. Shell color is bright golden wide midbody band of large patches and dark brown flammules, anterior tip marked with large white flammules; golden-tan and white base color overlaid with 21 extremely fine, hair-like, dark brown spiral lines. Shoulder and spire are white with large, evenly spaced, dark brown flammules. Spire flammules extend over edge of shoulder carina onto body whorl. Interior of aperture is white. This is a vormivorous species. Shells of this species vary in size from 16 to 65 mm. Like all species within the genus Conus, these snails are predatory and venomous.

Toxin Type: Conotoxin – O1 superfamily Br.7.6, Br. 7.8, Br 7.9, Br 7.10 and Br.7.11 precursors; O2 superfamily Br 7.1 and Br 7.5 precursors;

Conotoxin – A superfamily Br 1.4; O1 superfamily Br 7.7; O2 superfamily Br 7.1 to Br 7.4; (Morales-González et al., 2015; (http://www.conoserver. org/?page=list&table=protein&Organism_search%5B%5D=Conus%20 brunneus).

Alpha-conotoxin-like Br1.4 (http://www.uniprot.org/uniprot/P0C8U3);

Conus bullatus (Linnaeus, 1758)

Apertural view   **Abapertural view**

Common Name: Bubble cone

Geographical Distribution: Indian Ocean along the Mascarene Basin and Mauritius; in the Indo-West Pacific (the Philippines, New Caledonia)

Habitat: Intertidal habitats on coral rubble, coral, muddy sand, and gravel, down to depths of 240 m

Identifying Features: Shell of this species is medium-sized to large and moderately solid-to-solid. Last whorl is ovate to narrowly ovate. Outline is convex, less so or straight at adapical fourth and towards base. Aperture is distinctly wider at base than near shoulder. Shoulder is sub-angulate to angulate or slightly carinate. Spire is low and outline is either concave, with apex projecting from an otherwise almost flat spire, or straight. Teleoconch sutural ramps are flat and concave in late whorls, with 1 increasing to 2–4 spiral grooves that are weak in latest whorls. Last whorl is with a few weak narrow spiral grooves at base. Ground color is white, variably suffused with orange to violet. Last whorl is with spiral rows of orangish to reddish brown dots, dashes, bars and spots that alternate irregularly with white dots and often with triangular spots in some rows. Fine reddish brown axial lines may extend from shoulder to base. Shell size varies from 45 to 82 mm. Like all species within the genus Conus, these snails are predatory and venomous.

Toxin Type: Alpha-conotoxin BuIA (Azam et al., 2005).

Conus californicus **(Hinds, 1844) (=** *Californiconus californicus***)**

Apertural view   Abapertural view

Common Name: California cone

Geographical Distribution: Cooler, temperate waters of Eastern Pacific Ocean (from the Farallon Islands near San Francisco to Bahia Magdalena, in Baja California, Mexico)

Habitat: Rocky and sandy areas, in the intertidal zone and subtidally down to 30 m depth

Identifying Features: Shell of this species is distinguished by its grayish brown color and thick periostracum. It is round-shouldered with the aperture broader at the base. The spire is flat and the height of the shell ranges from 25 to 40 mm. This predatory cone snail hunts and eats marine worms, fish and mollusks. It is also a scavenger. Like all species within the genus Conus, these snails are predatory and venomous.

Toxin Type: Cal5a-5e, Cal14.2, Cal14.1a-c, Cal16.1, Cal16.1a, Cal16.1h, Cal16.2, Cal16.3; Cal6.3a & Cal6.3b (O- superfamily); Cal9.1, Cal9.2; Cal22e, Cal22f, Cal12.1.1–12.1.4; Cal12.2a-d, Cal9.1a, Cal14.1, Cal6.4, Cal5.1, Cal16.1 (Elliger et al., 2011).

Like all members of the genus, this species has a specialized venom apparatus, including a modified radular tooth, with which it injects paralyzing venom into its prey. Presence of putative peptides in material derived from the tooth lumen, and all of the more prominent species were also evident in the anterior venom duct. Radular teeth thus appear to be loaded with peptide toxins while they are still in the radular sac (Marshall et al., 2002; Elliger et al., 2011).

Conasprelloides cancellatus **(Hwass, C.H. in Bruguière, J.G., 1792)** *(= Conus austini)*

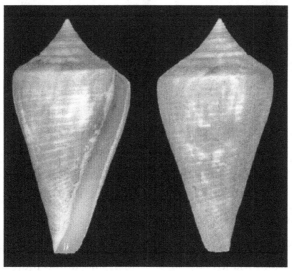

Apertural view   Abapertural view

Common Name: Austin's/Cancellate Cone

Geographical Distribution: Caribbean Sea, the Gulf of Mexico and the Lesser Antilles.

Habitat: Shallow and deep waters of 26–110 m

Identifying Features: Pear-shaped shell of this species is broad and angulated at the shoulder, contracted towards the base. Body whorl is closely sulfate throughout, the sulci striate and intervening ridges are rounded. Spire is carinate and concavely elevated. Its apex is acute and striate. Color of the shell is whitish, obscurely doubly banded with clouds of light chestnut. Maximum-recorded shell length is 80 mm. Like all species within the genus Conus, these snails are predatory and venomous.

Toxin Type: Kappa-conotoxin-like as14b (http://www.uniprot.org/uniprot/P0C6S3).

Conus capitaneus **(Linnaeus, 1758) (=** *Rhizoconus capitaneus***)**

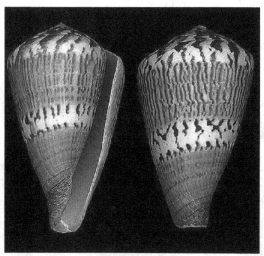

Apertural view   Abapertural view

Common Name: Captain cone

Geographical Distribution: Indian Ocean along Madagascar, the Mascarene Basin, Mauritius and Tanzania; and in the Indo-West Pacific (along Hawaii, Samoa, Tonga, Japan to Australia

Habitat: Intertidal and entire subtidal areas; near rocky shores, lower eulittoral, often under boulders.

Identifying Features: Its low spire is striate, flamed with chocolate and white. Body whorl is yellowish, or orange-brown, encircled by rows of chestnut dots, usually stained chocolate at the base. There is a central white band with chocolate hierogtyphic markings on either side, and a shoulder-band crossed by chocolate smaller longitudinal markings. Aperture is white. Size of an adult shell varies between 50 mm and 98 mm. Like all species within the genus Conus, these snails are predatory and venomous.

Toxin Type: Cap15a;

Conotoxin Cap15b (http://www.uniprot.org/uniprot/C8CK78);

T superfamily conotoxin Cap5.1 (http://www.uniprot.org/uniprot/S4UJF0);

Omega-conotoxin-like 1 (http://www.uniprot.org/uniprot/Q5K0C4);

Alpha-conotoxin-like Cp20.1 (www.hoelzel-biotech.com);

Alpha-conotoxin-like Cp20.2 (www.cusabio.com, 2007–2015);

Alpha-conotoxin-like Cp20.5 (www.cusabio.com, 2007–2015);

Alpha-conotoxin Cp20.3 (Conopeptide alpha-D-Cp) (www.cusabio.com. 2007–2015);

Conotoxin CaHr91 (www.cusabio.com. 2007–2015);

Conotoxin Cap15b (www.mybiosource.com/prods/Recombinant-Protein);

Conotoxin Cp1.1 (http://www.ebi.ac.uk/interpro/entry/IPR020242/proteins-matched?start=40).

Conus caracteristicus **(Fischer von Waldheim, 1807) (=** *Lithoconus caracteristicus*)

Abapertural view   Apertural view

Common Name: Characteristic cone

Geographical Distribution: E. Indian Ocean; Bay of Bengal, south to the Philippines and then north to Japan

Habitat: Subtidal sandy habitats to depths of around 30 m

Identifying Features: Shell of this species is heavy, and glossy and sides are straight. Body whorl is conical to broadly conical. Outline is convex below shoulder and is straight towards base. Shoulder is angulate to rounded. Spire is nearly flat and seldom slightly elevated. Early whorls form a small bluntly pointed cone. Body whorl is with strong spiral ridges anteriorly and is separated by wide grooves. Aperture is moderately wide, uniform in width and slightly flaring anteriorly. Outer lip is fairly thin. Ground color is glossy white or cream with three irregular spiral bands of broad, reddish brown blotches, which are widely interspaced below shoulder and on both sides of center. Shoulder is with alternating reddish-brown and is white blotches. Tip of spire is white. Aperture deep yellow. Shell size varies from 40 to 88 mm. Like all species within the genus Conus, these snails are predatory and venomous.

Toxin Type: S-superfamily conotoxin: ca8a (Liu et al., 2008);
    T-superfamily conotoxin: cr5a (Dovell, 2010).

Conus circumcisus **(Born, 1778)** (= *Pionoconus circumcises, Conus brazieri*)

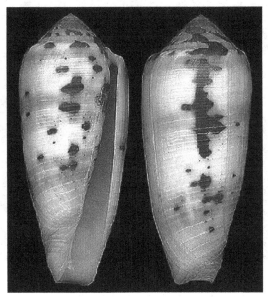

Apertural view   **Abapertural view**

Common Name: Circumcision cone, Circumcised Cone shell

Geographical Distribution: Pacific Ocean along Moluccas, Philippines, Marshall Islands, Solomon Islands and Vanuatu

Habitat: At depths of 4 to 200 m on sand, coral rubble, on lagoon pinnacles and in small caves on vertical cliffs

Identifying Features: Shell of this species is rather solid, with revolving striae throughout. Its color is whitish, tinged with pale rose-pink, with two broad, light yellowish brown bands, sprinkled here and there with a few very minute brown spots. Spire is conspicuously marked with dark brown blotches. Adults can grow up to 100 mm in length, although they will typically be smaller than this This species is a fish-hunting cone snail that immobilizes prey with a complex and powerful venom. Like all species within the genus Conus, these snails are venomous to humans.

Toxin Type: Superfamily conotoxin-Cr4.1 precursor (http://www. conoserver.org/?page=card&table=nucleicacid&id=1694).

Conus clerii **(Reeve, 1844)**

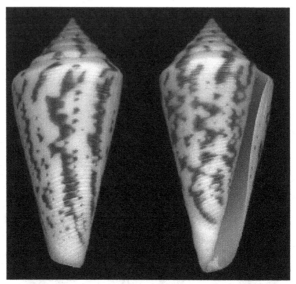

Abapertural view   Apertural view

Common Name: Clery's cone

Geographical Distribution: Endemic to Brazil; from the Cape of Sao Tome, Rio de Janeiro state to the border region with Uruguay

Habitat: Mainly found resting or crawling along the sea floor in fairly deep water (usually 10–100 m. deep); occasionally found in near-coastal waters and along beaches

Identifying Features: It is a small- to medium-sized species with a solid calcareous spiral-shaped shell, light in weight with a good-to-high gloss. Shell is low biconical in shape with a spire, which is straight to slightly concave. Whorl tops are flat to slightly concave with a few faint spiral ridges crossed by stronger axial ridges. Earliest whorls are weakly nodulose and are becoming smooth on latest whorls. Body whorl has variably angled shoulders (sometimes bluntly, sometimes very sharply), which are slightly convex-to-straight sides tapering to a narrow base. Shell aperture is uniformly wide and downward sloping below the shoulder, with a thin, sharp, straight to slightly convex outer lip. Columella is hidden from external view. Body whorl ground color is white, cream or pinkish tinted, with brown (often with an orange tint), irregular, vertically oriented with squiggle shaped markings.

Base is usually ground colored. Spire is white with variably sized, brown blotches, often radiating. Apex and shell interior are usually white or pinkish. Maximum-recorded shell length is 65 mm. Like all species within the genus Conus, these snails are venomous to humans.

Venom Characteristics: This species has a moderately potent neurotoxin with possible cardiotoxic factors. Sting effects vary with individuals, but usually include: localized very sharp pain (sometimes excruciating), inflammation, limited swelling, numbness, and mild cyanosis at the wound site. Some victims' may have mild-to-severe headache, nausea, vomiting, breathing difficulties and /or hearing impairment. Some cases of severe flaccid paralysis have been reportedly caused by envenomation (a "sting") by this species. Its venom may also be lethal to adult humans, and even lesser "stings" may trigger an anaphylactic reaction in some persons. Most "stings" inflicted on humans by this species, have occurred when the "pretty" shell (still containing the live animal) has been either stepped on or intentionally handled (e.g., picked up and examined) by an incautious person.

Conus consors **(G. B. Sowerby I, 1833)** (= *Pionoconus consors*)

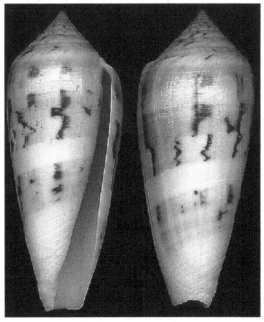

Apertural view   Abapertural view

Common Name: Singed cone

Geographical Distribution: Indo-West Pacific Region to the Marshall Islands, in Melanesia and along Queensland, Australia.

Habitat: Slightly subtidal to 200 m; in sand and silt.

Identifying Features: Shell of this species is medium sized and moderately solid to heavy. Body whorl is narrowly conical and outline is convex adapically with high gloss. About 10 to 12 low, rounded spiral ridges are seen above the base separated by shallow, weakly punctuate grooves. Rest of the body whorl is with numerous spiral and axial threads and growth lines. Shouldis broad, rounded and narrower than body whorl immediately anterior to it. Spire is moderately low, sharply pointed, sides straight. Aperture is fairly wide, and outer lip is slightly concave in the middle. Body whorl is pale yellow with two spiral bands above and below the center. Spiral band above the center is broader. Color of bands ranges from yellowish brown to dark brown. Usually indistinct dark brownish axial streaks are seen on the spiral band. Spire is yellowish to tan, almost with a few pale brown spots and streaks near sutures. Adult shell varies between 33 mm and 118 mm in size. These snails are predatory and venomous. As they are capable of "stinging" humans, live ones should not be handled.

Toxin Type: Conodipine isoforms (Robinson et al., 2014); Conotoxin CcTx (Terlau and Olivera, 2004); Alpha-conotoxin CnIA; Alpha-conotoxin CnIC (http://www.ebi.ac.uk/interpro/signature/PS60014/proteins-matched;jsessionid=72887C4196F6CFF05EE34D33FCB8F5C0).

Conus coronatus (Gmelin, 1791)

Apertural view   **Abapertural view**

Common Name: Crowned cone

Geographical Distribution: Throughout the Indo-Pacific, ranging from the east coast of Africa to French Polynesia and Hawaii

Habitat: Intertidally to about 10 m; semisheltered patches of sand, among seaweed, among rocks, on rough limestone benches, among rubble and coarse sand, on coral rocks (both with and without algal turf) as well as on fine sand and mudflats

Identifying Features: Shell of this species is medium to heavy in weight. Body whorl is with convex sides, sculptured with spiral grooves, varying from a few on the base to covering most of the body whorl. Shoulder is sub-angulate to angulate, coronate. Spire is low to medium height and straight sided. Whorls are with spiral grooves. Aperture is wider at base. Ground color is pinkish to violet, with pale bluish spiral bands below shoulder and center. Spiral bands are of variously sized brown and black markings on either side of subcentral band, overlaying the two solid color bands. Spire is similar to body color. Aperture pale gray, edges tinged dark violet. Maximum shell length is 47 mm. These snails are predatory and venomous. As they are capable of "stinging" humans, live ones should not be handled.

Toxin Type: Four-loop conotoxins (Duda and Remigio, 2008).

Conus crotchii **(Reeve, 1849)**

Apertural view   Abapertural view

Common Name: No common name

Geographical Distribution: Endemic to Boavista Island within the Cape Verde island group

Habitat: Isolated rocky reefs scattered along the sandy bays

Identifying Features: Protoconch of this species is paucispiral and whorl tops may be concave when viewed in cross section, with cords on the whorl tops that may be lost in middle spire whorls or persist thereafter. Shell has a shallow to moderately deep anal notch. Periostracum is smooth and thin, and operculum is small. Adults of the species grow to 30 mm in length. These snails are predatory and venomous. As they are capable of "stinging" humans, live ones should not be handled.

Toxin Type: All conopeptides detected belong to the A-, O1-, O2-, O3-, T- and D-superfamilies, which can block Ca2+ channels, inhibit K+ channels and act on nicotinic acetylcholine receptors (nAChRs) (Neves et al., 2013).

Conus dalli **(Stearns, 1873)** *(= Cylinder dalli)*

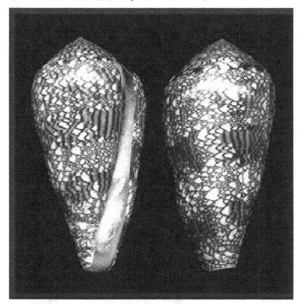

Apertural view   Abapertural view

Common Name: Dall's cone

Geographical Distribution: Eastern Pacific along the Galapagos Islands, and the Sea of Cortez to Panama.

Habitat: Sand pockets in rock terraces (to 50 m); intertidal areas in tide pools

Identifying Features: Spire of this species is indistinctly grooved. Body whorl is obscurely spirally ribbed below. Color of the shell is yellowish brown, with reddish brown longitudinal stripes, interrupted by four revolving bands of white spots, and occasional white spots on the darker surface. Interior of the aperture is rosy pink. Size of an adult shell varies between 32 mm and 80 mm. These snails are predatory and venomous. As they are capable of "stinging" humans, live ones should not be handled.

Toxin Type: Three novel conopeptides have been isolated and characterized; dal_C1011h, dal_C0910, and dal_C0805 g. dal_C1011h is a 27-residue hydrophobic conotoxin that belongs to O-superfamily, dal_C0910 is a 16-residue conotoxin that belongs to M-superfamily, and dal_C0805 g is a 12-residue linear conopeptide the belong to the Conorfamide family (Usama, 2005).

Conus delessertii **(Récluz, 1843)**

Apertural view   **Abapertural view**

Common Name: Sozon's cone

Geographical Distribution: Gulf of Mexico; South-East USA and Caribbean

Habitat: At depths of 15–198 m.

Identifying Features: Shell of this species is medium sized to large, light or moderately solid-to-solid. Last whorl is conical and outline is straight or slightly convex near shoulder. Shoulder is carinate and outline is straight or slightly concave. Spire is moderate to high and outline is straight to concave. Teloconch sutural ramp is concave to almost flat with many curved axial striae and 2–4 spiral striae separated by broader ribs. Last whorl is white with 3 broad light orange-tan to salmon spiral bands below shoulder and on each side of center. Numerous spiral rows of brown rectangular dots or dashes over entire whorl. Apex is pale brown. Spire and shoulder are white to pale salmon. Aperture is white. Maximum recorded shell length is 139 mm.

Toxin Type: Major peptides, de7a and de13a from the crude venom of this species (Aquilar et al., 2005).

Conus distans **(Hwass in Bruguière, 1792) (=***Conus (Rhombiconus) distans***)**

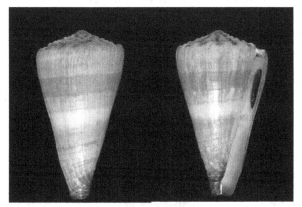

Abapertural view   Apertural view

Common Name: Distant cone, distantly lined cone

Geographical Distribution: Southern Natal to Red Sea and to Hawaii and French Polynesia;

Habitat: Intertidal and upper subtidal to about 40 m; from intertidal benches to the outer side of coral reefs; sand, rubble, algal turf, dead as well as living coral and caves; juveniles in greater depths

Identifying Features: Shell of this species is moderately large to large and solid to heavy. Last whorl is conical and outline is convex to strongly convex near shoulder, straight below. Shoulder is variably angulate and it is tuberculate in smaller adults (up to 75–85 mm in length). Spire is of low to moderate height and outline is concave to convex. Teleoconch sutural ramps are concave, with 2–3 faint spiral grooves in middle ramps, and finer spiral striation on later ramps. In subadults, last whorl is with broad spiral ribs at base, followed by about 5 widely set spiral grooves almost to adapical third with ribbons between. In large adults, surface is smooth except for broad but faint spiral ribs at base. Ground color is white. In smaller subadults, last whorl is with olive brown clouds and base is dark brown. During growth, clouded pattern is changing to a band on each side of center. Such bands are either progressively occupying entire last whorl (Pacific populations) or secondarily reduced (Indian Ocean populations). Bands usually become less olive and base is slightly lighter during growth. Teloconch spire is with blackish brown radial markings between tubercles (including large Indian Ocean shells without color bands on last whorl). Length of the shell varies from 65 to 137 mm. These snails are predatory and venomous. As they are capable of "stinging" humans, live ones should not be handled.

Toxin Type: An excitatory peptide, di16a, with 49 amino acids and 10 cysteine residues has been purified and characterized from the venom of this species.

Conus ebraeus (Linnaeus, 1758)

Apertural view   **Abapertural view**

Common Name: Black-and-white cone, Hebrew cone

Geographical Distribution: Tropical regions throughout the Indo-West and eastern Pacific, from the Red Sea to the shores of the Americas

Habitat: Intertidal and subtidal habitats to about 3 m, on sand, among or beneath dead corals and on coral reef and limestone platforms.

Identifying Features: Shell of this species is moderately small and heavy and is broadly conical with convex sides. Body whorl is with low spiral ridges and basically these often extend to center. Shoulder is roundly angled, and coronated with coronations. Spire is low and convex and is bluntly pointed. Spire whorls are weakly coronated and are with 4–5 heavy spiral ridges on top. Aperture is narrower posteriorly and outer lip is thick, sharp and convex. Ground color is white. Body whorl is with 3–4 spiral rows of black blotches between base and subshoulder. Blotches are squarish to more or less axially elongate. Shoulder is with large squarish black spots, alternating with light background. Spire is with irregular black blotches. Aperture is bluish. Height of the shell varies from 25 mm to 62 mm. These snails are predatory and venomous. As they are capable of "stinging" humans, live ones should not be handled.

Toxin Type: Four –loop conotoxin (Duda and Remigio, 2008);
I-superfamily conotoxins (http://www.ebi.ac.uk/interpro/entry/IPR020242/proteins-matched?start=40).

Conus eburneus **(Hwass in Bruguière, 1792)** (= *Lithoconus eburneus*)

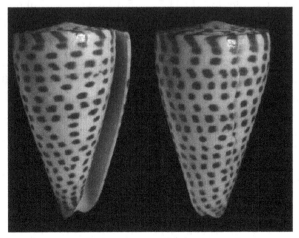

Apertural view   **Abapertural view**

Common Name: Ivory cone

Geographical Distribution: Indo-West Pacific from the coast of East Africa (along Madagascar and Chagos) to Australia, Polynesia and the Ryukyu Islands (but not along Hawaii)

Habitat: Shallow waters to 65 m; sand bottoms of subtidal reef flats; burrows in sand

Identifying Features: Shell of this species is medium sized and moderately solid with high gloss. Body whorl is conical and outline is convex at subshoulder and straight below. Base is truncate. Shoulder is rounded. Spire is low, pointed and sides are straight. Spire whorls are flat. Body whorl is with numerous weak spiral ribs and ribbons are seen on basal fourth to half. Aperture is narrow, and is slightly wider at base. Outer lip is thin and sharp. Ground color is white. Body whorl is with spiral rows of large, variably spaced reddish-brown squarish spots and rectangular bars of various sizes. Four light-yellow bands underlie spiral rows of spots below the shoulder, on both sides of center and on the base. Shoulder is with regular oblique brown spots, which are extending on to last spire whorl. Other spire whorls are with regular reddish-brown spots. Tip of spire is white. Aperture is white. Shell size varies from 35 to 79 mm. These snails are predatory and venomous. As they are capable of "stinging" humans, live ones should not be handled.

Toxin Type: Conotoxin Eb11.7 (http://www.ebi.ac.uk/interpro/entry/IPR020242/proteins-matched?start=40); δ (delta) conotoxin ErVIA (Aman et al., 10.1073/pnas.1424435112); T-superfamily conotoxins (Liu et al., 2012).

Eburne toxin, a powerful vasoactive protein has been isolated from the venom of this species. This toxin at concentrations above $3 \times 10 (-7)$ g/mL elicited a marked contractile response of rabbit aorta. The minimum lethal dose in the fish Rhodeus ocellatus was found to be 1 microgram/g body weight (Kobayashi et al., 1982).

Conus emaciatus (Reeve, 1849)

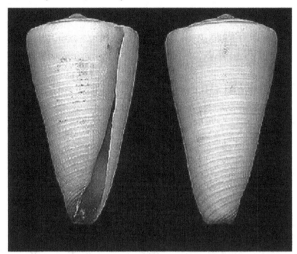

Apertural view   Abapertural view

Common Name: False virgin cone

Geographical Distribution: Red Sea and Central Indian Ocean to Polynesia

Habitat: Intertidal benches and shallow subtidal reef flats; sand bottoms, bare limestone or beachrock, and dead coral heads and rocks.

Identifying Features: Shell of this species is moderately small to moderately large and is moderately solid-to-solid. Last whorl is conical and outline is convex at adapical third, slightly concave at center and straight below. Shoulder is angulate to subangulate. Spire is usually low and its outline is concave to convex. Teleoconch sutural ramps are flat to slightly convex. Last whorl is with rather regularly spaced, often finely granulose spiral ribs with closely set spiral striae between from base to adapical third. Color of the shell is yellowish gray to orange and yellow. Last whorl is often slightly paler near center and at shoulder. Base and larval whorls are purplish blue. Aperture is white except for the basal area, occasionally tinged with violet deep within. Shell size varies from 30 to 69 mm.

These snails are predatory and venomous. As they are capable of "stinging" humans, live ones should not be handled.

Toxin Type: Kappa-conotoxin-like Em11.8; Kappa-conotoxin-like Em11.10; Conotoxin Em11.5 (http://www.ebi.ac.uk/interpro/entry/IPR020242).

Darioconus episcopatus **(Reeve, L.A., 1843) (=** *Conus episcopatus***)**

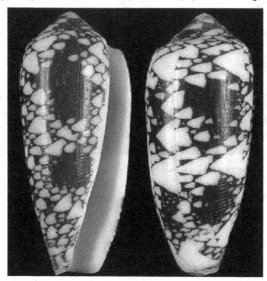

Apertural view and viceversa   Apertural view

Common Name: Bishop's cone; dignified/magnificent cone

Geographical Distribution: Indo-W Pacific

Habitat: Variety of lagoon and seaward reef habitats; in sand or rubble underneath rocks during the day and emerges at night to hunt gastropod mollusks

Identifying Features: Shell of this species is narrower, more cone-like shape and more erect spire with a rounded top. Its size varies from 40 to 115 mm in size. These snails are predatory and venomous. As they are capable of "stinging" humans, live ones should not be handled.

Toxin Type: Conotoxin Ep11.12 (http://www.ebi.ac.uk/interpro/entry/IPR020242/proteins-matched?start=40); α-Conotoxin EpI (a Novel

Sulfated Peptide which selectively targets neuronan Nicotinic acetylcholine receptors); I1-superfamily conotoxin, Ep11.1.

Vituliconus ferrugineus **(Hwass, C.H. in Bruguière, J.G., 1792)** **(= *Conus ferrugineus*)**

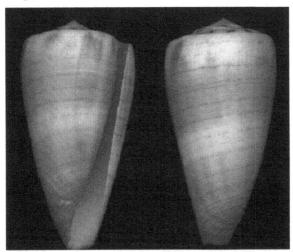

Apertural view   Abapertural view

Common Name: Not designated

Geographical Distribution: Pacific Ocean from Indonesia to the Marquesas islands, along Queensland, Australia and New Caledonia

Habitat: Sand which is often algae covered or under coral reefs; at intertidal depths of between 5 and 50 m

Identifying Features: Thin shell of this species has a depressed carinate and striate spire, which is yellowish, maculated with brown. Body whorl is striated below, yellowish, with two series of longitudinal forked and irregular dark brown markings, interrupted in the middle and at the base. There are traces of distant narrow brown revolving lines. Aperture is white. Size of an adult shell varies between 40 mm and 80 mm. It is known to feed on polychaetes and has a planktonic larval development. These snails are predatory and venomous. As they are capable of "stinging" humans, live ones should not be handled.

Toxin Type: Novel P and T-Superfamilies' of conotoxin; Alpha/kappa-cono-toxin-like fe14.1; Fer B; Alpha/kappa-conotoxin-like fe14.2 (http://www.uniprot.org/uniprot/Q0N4U3); Conotoxin fe11.1; Conotoxin fe11.1 (http://www.ebi.ac.uk/interpro/entry/IPR020242/proteins-matched?start=40); O gene superfamily of conotoxins-fe6.1 and fe6.2;

Fer_B03 g, Fer_B05p, Fer_D04ij, Fer_D05r, F06k, Fer_F07i (Pak, 2014).

Conus figulinus (Linnaeus, 1758)

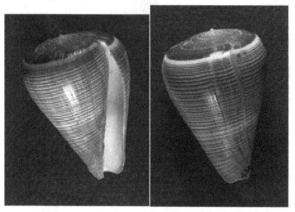

Apertural view   Abapertural view

Common Name: Fig cone

Geographical Distribution: Indian Ocean along Madagascar, the Mascarene Basin, Mauritius and Tanzania; in the Indo-West Pacific.

Habitat: Intertidal and uppermost subtidal (up to 25 m); in semi sheltered or protected sites, living on fine to very fine sand of flats, often among vegetation

Identifying Features: Shell of this species is medium-sized to large and solid to heavy. Last whorl is ventricosely conical to broadly ventricosely conical or slightly pyriform. Outline is convex adapically and straight or slightly concave towards bas. Left side is consistently sigmoid. Shoulder is rounded. Spire is usually low and its outline is variably sigmoid or concave. Teleoconch sutural ramps are flat to slightly convex, with many spiral striae. Last whorl is with variably pronounced and spaced spiral grooves on basal third, separating ribs and ribbons. Ground color varies from yellowish or orangish cream through reddish to grayish or blackish brown. Last whorl is with variably spaced, solid or occasionally dashed or dotted spiral lines of brown or black. These lines are usually absent from a narrow band below shoulder edge. Subshoulder band may contrast in color from adjacent area of last whorl, ranging from yellow to dark reddish brown. Teleoconch sutural ramps are orange-brown to blackish brown, darker than last whorl. Aperture is white or bluish white. Size of an adult shell varies between 45 mm and 135 mm. These snails are predatory and venomous. As they are capable of "stinging" humans, live ones should not be handled. It probably feeds on polychaetes.

Toxin Type: M-Superfamily conotoxins- Fi3a, Fi3b, and Fi3c; T-Superfamily Fi5a; contryphans fib, fic, and fid (Rajesh, 2015);

Crude extract e\of this species has shown haemolytic activity (Saravanan, et al., 2009).

Conus flavidus **(Lamarck, 1810)**

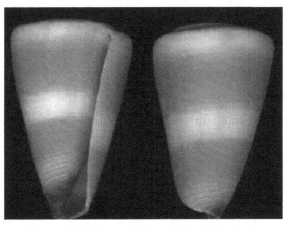

Apertural view   Abapertural view

Common Name: Yellow Pacific cone, Golden yellow cone

Geographical Distribution: Entire Indo-Pacific

Habitat: Intertidal benches; shallow subtidal reef flats to about 20 m; from inshore habitats to the reef rim; small sand-filled depressions; reef limestone or beachrock with or seldom without algal turf; coral rubble with or without sand; dead coral heads or rocks.

Identifying Features: Shell of this species is medium-sized to moderately large and moderately solid to moderately heavy. Last whorl is conical to broadly conical, sometimes slightly ventricosely conical and its outline is variably convex at adapical fourth or third and almost straight below, often slightly concave centrally in large specimens. Shoulder is angulate to subangulate. Spire is usually low and outline is slightly concave to slightly convex. Teleoconch sutural ramps are flat to slightly concave and its later ramps are with numerous spiral striae, with 3–6 spiral grooves, or with intermediate sculpture. Last whorl is with distinct to obsolete spiral ribs basally. These ribs are weaker adapically. Color of the shell is light yellowish to orange or pinkish brown, occasionally brownish green. Last whorl is with a pale or white spiral band at or closely below center and at shoulder. Base is purplish blue. Larval whorls change color from white to purple during metamorphosis. Postnuclear sutural ramps vary from white to color of last whorl. Aperture is violet to purplish blue, with pale bands near center and below shoulder. This species feeds on sedentary poly-chaetes of the families Terebellidae, Maldanidae and Capitellidae, rarely consuming enteropneusts. Diet composition varies depending on habitat and locality. Shell size varies from 35 to 75 mm. These snails are preda-tory and venomous. As they are capable of "stinging" humans, live ones should not be handled.

Toxin Type: A-conotoxin peptide Fla1.7a (http://www.uniprot.org/uniprot/U3KZV0);

O1-conotoxin peptide Fla-17 (http://www.uniprot.org/uniprot/U3KZV2);

Q-conotoxin peptide (http://www.uniprot.org/uniprot/V5 V8A2);

I2-conotoxin peptide Fla11.1 (http://www.ebi.ac.uk/interpro/entry/IPR020242/proteins-matched?start=40);

M superfamily Fla-15; Fla3.2 to Fla3.5;

A superfamily Fla1.1, Fla1.2, Fla1.3, Fla1.6, Fla1.7, Fla1.8, Fla6.1;

Superfamily -Fla16.1 to 16.8; Fla6.14 to Fla6.17;

O1 superfamily Fla6.3 to Fla6.6;

O2 superfamily Fla6.8;

(http://www.conoserver.org/?page=list&table=protein&Organism_search%5B%5D=Conus%20flavidus – Cono Server).

Conus fulmen (Reeve, 1843)

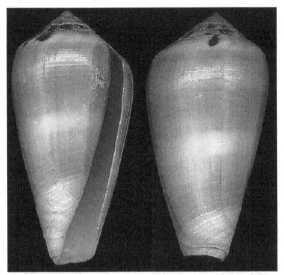

Apertural view   **Abapertural view**

Common Name: Thunderbolt cone

Geographical Distribution: Japan to Taiwan

Habitat: At depths of 35–100 m.

Identifying Features: Shell of this species is medium-sized to moderately large, and moderately solid-to-solid. Last whorl is ventricosely conical and outline is convex adapically, less so or straight below. Shoulder is subangulate to rounded. Spire is of low to moderate height and its outline is straight to slightly convex. Larval shell is of about 2.75 whorls with a maximum diameter of 0.9 mm. About first 4 postnuclear whorls are tuberculate. Teleoconch sutural ramps are convex to almost flat, with 2 increasing to 4 spiral grooves in early whorls, and many spiral striae are seen in later whorls. Last whorl is with wrinkled spiral ribs abapically, often followed by spiral threads. Color of the shell is violet blending with white. Last whorl is encircled with continuous, broad or narrow, violet to tan bands above and below a narrow light band at center. Solid or dotted brown spiral lines and dark brown axial blotches and flames vary in number, arrangement and prominence. Larval whorls are orangish red.

Teleoconch sutural ramps are with variably broad reddish to blackish brown radial blotches. Aperture is white or pale violet. Shell size varies from 45 to 80 mm. These snails are predatory and venomous. As they are capable of "stinging" humans, live ones should not be handled.

Toxin Type: ω-conotoxin – FVIA (Lee et al., 2010);

Though Conotoxin (CTx), CTx-FVIA and CTx-MVIIA (Ziconotide*) depressed arterial blood pressure immediately after administration, pressure recovered faster and to a greater degree after CTx-FVIA administration (Lee et al., 2010);

Ziconotide (CTx-MVIIA; Prialt®) is the first conotoxin-derived drug to be approved for the treatment of refractory pain by the U.S. Food and Drug Administration (FDA).

***Pionoconus gauguini* (Richard, G., and Salvat, B., 1973) (= *Conus gauguini*)**

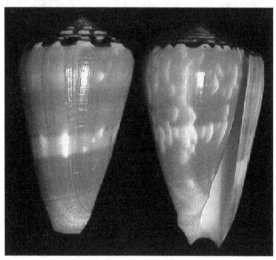

Abapertural view   Apertural view

Common Name: Gauguin's cone

Geographical Distribution: Pacific Ocean along the Marquesas and Tahiti

Habitat: Found buried in sand at the base of rubble slopes under large slabs of rock at depths of 18 to 35 m.

Identifying Features: Size of an adult shell varies between 56 mm and 93 mm. These snails are predatory and venomous. As they are capable of "stinging" humans, live ones should not be handled. Detailed description is wanting.

Toxin Type: Conotoxins- Ga1.1, Ga1.1 precursor, Ga1.2 and Ga1.2 precursor (Conoserver- http://www.conoserver.org/?page=list&table=protein &Organism_search%5B%5D=Conus%20 gauguini).

Conus generalis **(Linné, 1767)**

Apertural view   **Abapertural view**

Common Name: General cone

Geographical Distribution: Red Sea, in the Indian Ocean along Madagascar, Mauritius and Tanzania; in the Indo-West Pacific along Indonesia and the Philippines and from North-west Australia to French Polynesia and the Ryukyu Islands; in the Central Indian Ocean along the Maldives.

Habitat: Sand, muddy sand, coral rubble in subtidal habitats; At depths of 0–240 m.

Identifying Features: Spire of this species is rather plane, with a characteristic, small, acuminate, raised apex. Color of the shell is orange-brown to chocolate and is irregularly white-banded at the shoulder, in the middle, and at the base. These two or three bands are overlaid with zigzag or irregular chocolate-colored markings. Aperture is white. Size of an adult shell varies between 45 mm and 105 mm. It feeds on polychaetes. These snails are predatory and venomous. As they are capable of "stinging" humans, live ones should not be handled

Toxin Type: Two conotoxins identified from this species are the first natural peptides, and they are classified as members of the O2-superfamily (Xu et al., 2015).

Conus gladiator **(Broderip and Sowerby, 1833)** (= *Gladioconus gladiator*)

Apertural view   **Abapertural view**

Common Name: Gladiator cone

Geographical Distribution: Pacific Ocean along the Galapagos Islands and from the Sea of Cortez to Peru.

Habitat: Rocky areas sometimes covered in algae at depths between 0 and 5 m

Identifying Features: Shell spire of this species is rather depressed, tuberculate and striate. Color of the shell is chocolate-brown, variegated with white, disposed in longitudinal streaks, with an irregular white band, and more or less distinct revolving lines of darker brown. Interior is white or tinged with chocolate. Epidermis is fibrous. Size of an adult shell varies between 26 mm and 48 mm. These snails are predatory and venomous. As they are capable of "stinging" humans, live ones should not be handled.

Toxin Type: Conophan gld-V (Gamma-hydroxyconophan gld-V).

Conus infinitus **(Rolán, 1990)** (*=Africonus infinitus*)

Apertural view   Abapertural view

Common Name: Not designated

Geographical Distribution: Endemic to the Cape Verde Islands in the West African Region.

Habitat: Not reported

Identifying Features: Shell of this species has a shallow to moderately deep anal notch. Periastracum is smooth and thin, and the operculum is small. This species is vermivorous, for example, it preys on marine worms. Size of an adult shell varies between 12 mm and 25 mm. These snails are predatory and venomous. As they are capable of "stinging" humans, live ones should not be handled.

Toxin Type: Megalin-like lipoprotein (http://www.uniprot.org/uniprot/ Q0D2Y8).

Conus inscriptus (Röckel, 1979)

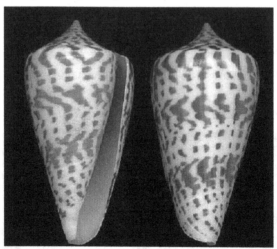

Apertural view   Abapertural view

Common Name: Not designated

Geographical Distribution: Indian Ocean, from Natal to Red Sea and to W. Thailand

Habitat: In 5–85 m depths

Identifying Features: Shell of this species is medium-sized to moderately large, and is usually moderately solid-to-solid. Shells from Mascarenes, Aden and Red Sea are smaller than those from other areas. Last whorl is ventricosely conical-to-conical and outline is convex at adapical fourth to half, usually straight below. Left side is sometimes concave near base and convex at adapical two-thirds. Shoulder is angulate to subangulate. Spire is of low to moderate height and highest in shells from Somalia to Mozambique. Its outline is concave to straight, most frequently straight and sometimes with stepped whorls in E. African shells. Teleoconch sutural ramps are flat to moderately concave, with 1 increasing to 3–8 spiral grooves. Last whorl is with widely spaced, weak to pronounced spiral grooves separated by ribbons on basal third to two-thirds. Ground color is white to beige or pale orange. Last whorl is with spiral rows of brown or orange dots, spots, bars or axial streaks. Subshoulder band is usually less prominent than anterior bands, sometimes absent. Aperture is white, beige to orange, pinkish or bluish violet, or pink. Shell size varies from 40 to 74 mm. These snails are predatory and venomous. As they are capable of "stinging" humans, live ones should not be handled.

Toxin Type: Contryphan-In (Gowd et al., 2005).

Conus judaeus **(Bergh, 1895)** *(= Virroconus judaeus)*

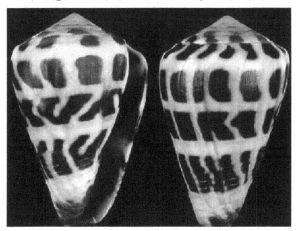

Apertural view   Abapertural view

Common Name: Not designated

Geographical Distribution: Philippines, the Seychelles, Okinawa in Japan and around Christmas Island

Habitat: On sand between algae or under coral rocks, on coral rubble, rock platforms and reef limestone from intertidal to 150 m

Identifying Features: Shell size is 32 x 21 mm². It feeds on polychaetes. Once mature it can reach a size ranging from 21 to 45 mm. These snails are predatory and venomous. As they are capable of "stinging" humans, live ones should not be handled.

Toxin Type: Four-loop conotoxin (Duda and Remigio, 2008);
  J6.1, J6.1 precursor, J6.2 and J6.2 precursor;
  (http://www.conoserver.org/?page=list&table=protein&Organ
ism_search%5B%5D=Conus%20judaeus).

Conus kinoshitai **(Kuroda, 1956)** (*=Afonsoconus kinoshitai*)

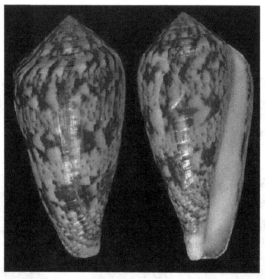

Abapertural view   Apertural view

Common Name: Kinoshita's cone

Geographical Distribution: South China Sea and in the Pacific Ocean
from the Philippines to the Solomons. There are also records in the Indian
Ocean from Mozambique, Madagascar, and Réunion

Habitat: Depths of 0–205 m

Identifying Features: Size of an adult shell varies between 40 mm and
94 mm. These snails are predatory and venomous. As they are capable of
"stinging" humans, live ones should not be handled.

Toxin Type: Alpha-conotoxin superfamily protein (http://www.uniprot.
org/uniprot/D4HRK8);
  Mu-conotoxin KIIIA (http://www.uniprot.org/uniprot/P0C195).

Conus lentiginosus (Reeve, 1844)

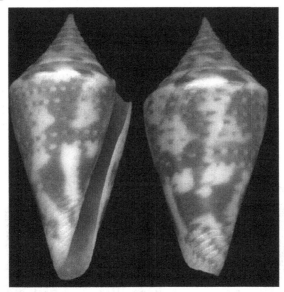

Apertural view   Abapertural view

Common Name: Brown-flamed cone

Geographical Distribution: Endemic to the west coast of India; possibly along a coastline of approx. 1,400 km.

Habitat: Shallow, subtidal sandy habitats

Identifying Features: Shell of this species is moderately small to medium-sized, and moderately light to moderately solid. Last whorl is conical, ventricosely conical or slightly pyriform. Outline is convex at adapical two-thirds to three-fourths and straight or concave below. Left side is consistently sigmoid and right side sometimes is almost straight. Shoulder is angulate. Spire is of moderate height and outline is concave or sigmoid. Teleoconch sutural ramps are flat to slightly concave, with 1 increasing to 1–3 spiral grooves. Last whorl is with axially striate spiral grooves on basal third, separated by ribs anteriorly and by ribbons posteriorly. Ground color is white, variably tinged with violet. Spiral rows of brown dots and dashes extend from base to shoulder, varying in number and arrangement. Larval whorls are white. Teleoconch sutural ramps are with brown radial blotches. Aperture is white, tinged with violet deep within. Shell size varies from 29 to 38 mm. These snails are predatory and venomous. As they are capable of "stinging" humans, live ones should not be handled.

Toxin Type: The crude venom of this species exhibited hemolytic activity on chicken erythrocytes, which was estimated as 8 HU. Further, this venom also exhibited neurostimulatory response on mice brain AChE activity. Inhibitory effect on AChE activity was found ranging between 23% and 397% (Kumar et al., 2014).

Conus leopardus **(Röding, 1798) (=** *Lithoconus leopardus***)**

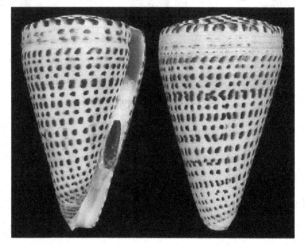

Apertural view   Abapertural view

Common Name: Leopard cone

Geographical Distribution: Indian Ocean along Aldabra, Chagos, Madagascar, the Mascarene basin, Mauritius and Tanzania; in the entire Indo-Pacific

Habitat: At depths of between 2 and 45 m in sand and rubble, at times with algae, on reef flats and typically in shallow bays.

Identifying Features: Shell of this species is large and heavy. Body whorl is usually conical and outline is almost straight. Shoulder is broad, angulate, and occasionally subangulate. Spire is very low or flat, sometimes moderate height and outline is slightly concave to slightly convex. Body whorl is with weak spiral ribs above base, obsolete in larger specimens. Ground color is white to cream and body whorl is with spiral rows of rounded or squarish, reddish brown spots or sometimes short axial streaks from base to shoulder, these sometimes in alternating large and small series. Shoulder and spire are with many narrow revolving dark brown broader lines on white. Aperture is relatively narrow, slightly wider anteriorly. This species is vermiverous and feeds exclusively on hemichordates. Size of an adult shell varies between 50 mm and 222 mm. These snails are predatory and venomous. As they are capable of "stinging" humans, live ones should not be handled.

Toxin Type: Conotoxins Leo-T1 and Leo-T2 (cusabio.com.);

Alpha-conotoxin-like (http://www.uniprot.org/uniprot/Q6PTD3);

Recombinant Conus leopardus Conotoxin Leo-O3 (Biocompare) (MyBioSource.com);

T-superfamily conotoxins, Lp5.1 and Lp5.2 (Chen et al., 2006).

Conus limpusi **(Röckel and Korn, 1990)** (*=Asprella limpusi*)

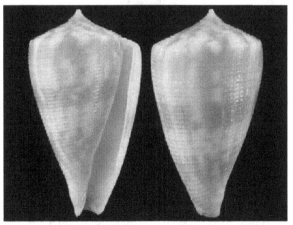

Apertural view   Abapertural view

Common Name: Limpus's cone

Geographical Distribution: Queensland, Australia

Habitat: Found on mud shell substrate at depths between 80 m and 225 m.

Identifying Features: Shell size varies from 30 to 53 mm. Once mature, it can reach a size ranging from 30 mm to 45 mm. These snails are predatory and venomous. As they are capable of "stinging" humans, live ones should not be handled.

Toxin Type: α-conotoxin LsIA;
    Synthetic LsIA was found to be a potent antagonist of α3β2, α3α5β2 and α7 nAChRs, with half-maximal inhibitory concentrations of 10, 31 and 10 nM, respectively (Inserra et al., 2013).

Conus longurionis (Kiener, 1845)

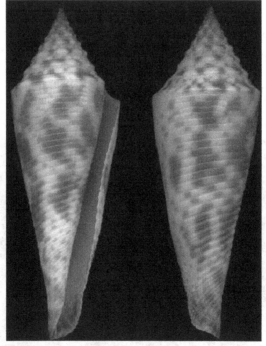

Apertural view   Abapertural view

Common Name: Cone snail

Geographical Distribution: Found in two separate regions of the Indo-Pacific: (1) Tanzania and northern Mozambique, and (2) India, Sri Lanka, western Thailand and Malaysia

Habitat: Buried in the sand of subtidal fringe; at depths down to 75 m where the habitat is primarily silt.

Identifying Features: Shell of this species is moderately small to medium sized and light. Body whorl is usually narrowly conical and outline is nearly straight. Shoulder is subangulate and exhalent notch is deep. Spire is hi and its outline is almost straight. First 3–10 post nuclear whorls are tuberculate and distinct. Sutures are deep and wide. Body whorl is with regularly spaced and axial striate spiral grooves and ribbons. Aperture is long, narrow and slightly wide anteriorly. Ground color is light brown. Body whorl is with spiral rows of regular large brown dots on ribbons, partly fusing into irregularly sized axial flecks that cluster into spiral bands above and below center. A weaker spiral band is seen around shoulder. Aperture is pale lavender. Adults of the species will grow to 46 mm. These snails are predatory and venomous. As they are capable of "stinging" humans, live ones should not be handled

Toxin Type: O-superfamily of conotoxins, having a molecular mass of 2589.00 Da and 2781.13 Da, respectively. (Lebbe, et al., 2014); α-conotoxin Lo1a, an 18-amino acid peptide which is active on nAChRs (Lebbe et al., 2014; http://www.uniprot.org/uniprot/X1WB75).

Conus lorosii (Kiener, 1845)
Image not available

Common Name: Vermivore cone

Distribution: No data

Habitat: At a depth of 50 m

Identifying Features: Shell of this species is medium sized to large and solid to heavy. Body whorl is usually ventricosely conical and outline is convex adapically and straight toward base. Shoulder is subangulate to rounded. Spire is low to moderate height and its outline is variably concave. Basal third of last whorl is with variably spaced spiral grooves separating ribs and ribbons. Ground color is gray mixed with pale blue and brown and violet in some. Colors are arranged in blending spiral and axial zones. Body whorl is with contrasting light narrow spiral bands at shoulder and below center. Shoulder band is always present but often it is very narrow and inconspicuous. Solid or interrupted reddish to blackish brown lines occur infrequently on body whorl. Aperture is usually white to bluish-white, sometimes reddish-brown. It feeds mainly on polychaetes. Maximum size of the shell is 97 mm. These snails are predatory and venomous. As they are capable of "stinging" humans, live ones should not be handled.

Nothing much is known about the biology of this species

Toxin Type: Conotoxin Lo959 (Gowd et al., 2005; Ramasamy and Manikandan, 2011; Sabareesh et al., 2006);

Crude extracts of this species have shown hemolytic activity (Janeena et al., 2015);

Contryphan A.

Conus lynceus **(G. B. Sowerby II, 1858) (= *Graphiconus ynceus*)**

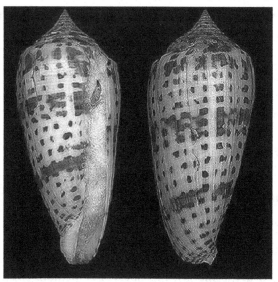

Apertural view   Abapertural view

Common Name: Lynceus cone

Geographical Distribution: Indo-Pacific along Taiwan, the Philippines, Java, Solomon Islands, Queensland, Australia

Habitat: At depths of between 20 and 50 m

Identifying Features: Shell of this species is somewhat swollen, distantly sulcate below, otherwise smooth. Color of the shell is white with encircled by chestnut spots, clouds, and oblique and triangular markings. It has a very pointed, maculated spire. Size of an adult shell varies between 50 mm and 89 m. These snails are predatory and venomous. As they are capable of "stinging" humans, live ones should not be handled.

Toxin Type: Iota-conotoxin-like L11.5 (http://www.uniprot.org/uniprot/P0C612);

Conantokin-L (http://www.uniprot.org/uniprot/P0C612).

Conus miles (Linnaeus, 1758)

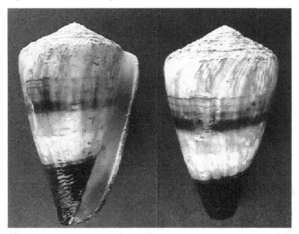

Apertural view   **Abapertural view**

Common Name: Soldier cone

Geographical Distribution: Aldabra, Chagos, Madagascar, Mascarene Basin, Mauritius, Mozambique, the Red Sea and Tanzania; entire Indo-Pacific.

Habitat: Intertidal; more commonly found in the subtidal zone up to a depth of 50 m; bays, intertidal reef, subtidal reef, sand, gravel, reef limestone, beachrock and lagoons.

Identifying Features: Shell of this species is moderately large, solid and glossy. Outline is broadly conical. Shoulder is angulate, broad, and weakly undulate. Spire is of moderate height and is bluntly pointed. Outline is straight. Body whorl is smooth except widely spaced spiral ribs on basal third and few spiral threads in between. Aperture is moderately wide and uniform in width. Outer lip is thin and sharp. Ground color is dull white. Body whorl is with a narrower light-brown spiral band above center. There is also a wider deep solid brown band on basal fourth to third and remaining areas are clouded with lighter brown crossed by well separated fine orange axial lines that extend to shoulder ramp. Spire is white and is covered with fine orange brown axial lines. Aperture is translucent. It is a vermivorous species hunting mainly on polychaetes. An adult shell varies between 50 mm and 136 mm. These snails are predatory and venomous. As they are capable of "stinging" humans, live ones should not be handled.

Toxin Type: O-superfamily peptides (Luo et al., 2007);

Conotoxin Mi11.1 (http://www.ebi.ac.uk/interpro/entry/IPR020242/ proteins-matched?start=40);

Alpha-conotoxin-like Ml20.2 (http://www.cusabio.com/Recombinant-Protein/Recombinant-Conus-miles–Alpha-conotoxin-like-Ml202–211015. html).

Conus miliaris (Hwass, 1792)

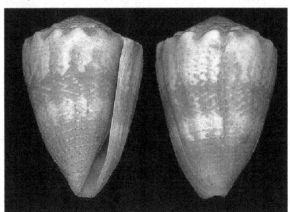

Apertural view   Abapertural view

Common Name: Thousand-spot cone

Geographical Distribution: Tropical to subtropical seas; from the Red Sea and eastern shores of Africa in the western Indian Ocean (Aldabra, Chagos, Kenya, Madagascar, the Mascarene Basin, Mauritius, Mozambique and

Tanzania) to Easter Island and Sala *y* Gómez in the south-eastern Pacific (but not along the Galapagos Islands, the Marquesas Islands and Hawaii).

Habitat: Shallow water environments

Identifying Features: Spire of this species is more or less raised, striate or sometimes nearly smooth, with or without tubercles. Body whorl is striate which is usually grannlous towards the base, and sometimes throughout. Color of the shell is yellowish or light chestnut or grayish, variously clouded with darker chestnut or olive, often irregularly light-banded at the middle, and below the spire, and encircled with chestnut spots on the striae. Interior is chocolate, with a central white band. There is considerable variation in the height and coronation of the spire, as well as in the color and pattern of the markings. Size of an adult shell varies between 12 mm and 43 mm. These snails are predatory and venomous. As they are capable of "stinging" humans, live ones should not be handled.

Toxin Type: O-superfamily conotoxin (http://www.uniprot.org/uniprot/C4NU44);

Four-loop conotoxins (Duda and Remigio, 2008).

Conus monile (Hwass in Bruguière, 1792)

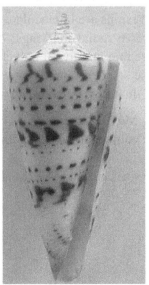

Apertural view

Common Name: Necklace cone

Geographical Distribution: India and Sri Lanka to W. Thailand; probably also Gulf of Oman.

Habitat: Intertidal to about 20 m; on sand bottom with patches of clay and shell rubbles

Identifying Features: Shell of this species is medium-sized to large, and moderately solid to heavy. Last whorl is conical and outline is variably convex at adapical fourth to third and straight below. Shoulder is angulate to carinate. Spire is of low to moderate height and its outline is deeply concave with a projecting conical apex of 5–7 postnuclear whorls. Maximum diameter of larval shell is about 0.9 mm. First 7–9 postnuclear whorls are tuberculate. Teleoconch sutural ramps are flat to concave, with obsolete spiral striae. Last whorl is with weak to obsolete spiral ribs at base. Ground color is white or cream. Last whorl is suffused or spirally banded with pale orange or pink. Spiral rows are of brown dots, dashes and variously shaped spots extend from base to shoulder but vary in number and arrangement, often concentrated at both sides of center. Sometimes dark markings fuse into axial flames or blotches. Base is pale orange or brown. Larval whorls are grayish beige. Early postnuclear sutural ramps are immaculate and late ramps are with a varying number of brown radial markings. Aperture is white. It is a vermivorous species feeding mainly on polychaetes. Shell size varies from 45 to 95 mm. These snails are predatory and venomous. As they are capable of "stinging" humans, live ones should not be handled.

Toxin Type: Conotoxin Mo1659 (a novel 13-residue acyclic peptide targeting noninactivating voltage-dependent potassium channels) (Gowd et al., 2005);
    Conotoxin Mo1274 (Ramasamy and Manikandan, 2011; Dovell, 2010).

Conus musicus **(Hwass in Bruguière, 1792) (=** *Harmoniconus musicus***)**

Abapertural view   Apertural view

Common Name: Music cone

Geographical Distribution: Red Sea and in the Indian Ocean along Aldabra, Chagos, Madagascar, Mozambique and Tanzania; in the Central Indian Ocean (along Sri Lanka and the Maldives) to the Marshall Islands and Fiji, Ryukyu Islands to West and East Australia

Habitat: At depths of 1–20 m where it occurs on intertidal and subtidal reef flats, the reef rim and other rocky surfaces such as dead coral, coral heads and pinnacles and in crevices

Identifying Features: Shell of this species is small, light to moderately light. Last whorl is conical or ventricosely conical to broadly ventricosely conical. Outline is faintly to distinctly convex at adapical half and usually straight below. Aperture may have a transverse ridge at center. Shoulder is angulate to occasionally rounded, weakly to distinctly tuberculate. Spire is of low to moderate height and its outline is slightly concave to slightly convex. Teleoconch sutural ramps are flat, in later whorls with 2 increasing to 3–4 spiral grooves. Last whorl is with weak to distinct and granulose spiral ribs are at base. Ground color is white to pale gray. Last whorl is with a gray, orange or reddish brown spiral band on each side of center. Bands are occasionally obsolete or fusing into a single basal color zone. Spiral rows of brown dots and dashes extend from base to shoulder, varying in number and arrangement. Dark dots may alternate with white dashes or dots. Base and basal part of columella are dark bluish violet. Later sutural ramps are crossed by brown markings between shoulder tubercles. Aperture is pale violet to dark bluish violet, usually with a ground-color band at center and below shoulder. Shell size varies from 14 to 30 mm. This species largely feeds on polychaetes of the families Nereidae and Eunicidae and they are only active at night. These snails are predatory and venomous. As they are capable of "stinging" humans, live ones should not be handled.

Toxin Type: A novel omega conotoxin belonging to the 'four-loop' structural class (Raju, 2007);

Conotoxin sequence 349,347,271,265,263,261,259,257,255 and 1251(http://www.conoserver.org/index.php?page=list&table=protein&Organism_search%5B%5D=Conus+musicus&sort_by=IDtext).

Conus mustelinus (Hwass, in Bruguière, 1792)

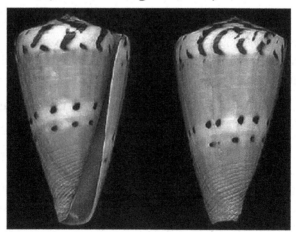

Apertural view   Abapertural view

Common Name: Weasel cone

Geographical Distribution: Indian ocean from the Chagos Atoll to Western Australia; in the Pacific Ocean, from Japan to Philippines, Eastern Australia and Fiji.

Habitat: Shallow waters of the subtidal on reefs and on sand beneath dead coral and rocks or in crevices

Identifying Features: In the shell of this species, spire is low and striate and is flamed with chocolate and white. Body whorl is yellowish, or orange-brown, encircled by rows of chestnut dots, usually stained chocolate at the base. There is a central white band, with chocolate hierogtyphic markings on either side, and a shoulder band, crossed by chocolate smaller longitudinal markings. Border markings of the bands are reduced to spots. Aperture has a chocolate color with a white ban. An adult shell varies between 40 mm and 107 mm. This species feeds on polychaetes. These snails are predatory and venomous. As they are capable of "stinging" humans, live ones should not be handled.

Toxin Type: D-superfamily conotoxins Ms20.1 to Ms20.5 (http://www.conoserver.org/?page=list&table=protein&Organism_search%5B%5D=Conus%20 mustelinus- Cono Server);

Alpha-conotoxin-like Ms20.5 (http://www.uniprot.org/uniprot/P0CE30);

Alpha-conotoxin Ms20.3 (http://www.uniprot.org/uniprot/C3 VVN5);

Alpha-conotoxin-like Ms20.2 (http://www.cusabio.com/Recombinant-Protein/Recombinant-Conus-mustelinus–Alpha-conotoxin-like-Ms202–212221.html).

Conus natalis **(G. B. Sowerby II, 1858)**

Abapertural view   **Apertural view**

Common Name: Natal textile cone

Geographical Distribution: Endemic to South Africa, and its range extends from Trafalgar to East-London

Habitat: Shallow waters between 2–55 m; muddy sand under large rocks

Identifying Features: Shell of this species is conical to cylindrical in shape with a conic spire and angular to subangulate shoulders. Whorl tops have an enlarged ridge in the center. Body whorl is fairly smooth and is ornamented with spiral lines of minute tents. This species preys on other gastropods and are preyed on by other snails. These snails are predatory and venomous. As they are capable of "stinging" humans, live ones should not be handled.

Toxin Type: Conotoxin nt3a (http://www.uniprot.org/uniprot/P0DMB3).

Conus nobilis **(Linnaeus, 1758) (=** *Eugeniconus friedae***)**

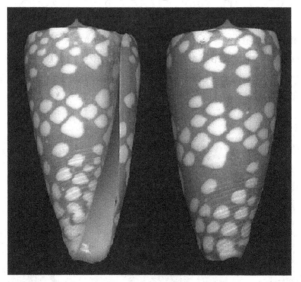

         Apertural view   **Abapertural view**

Common Name: Noble cone

Geographical Distribution: Indo-Pacific from Sri Lanka, the Andaman Islands and Nicobar Islands along Sumatra and Java to Timor; along the Marquesas Islands

Habitat: Sublittoral and deeper waters; on sand bottom with Foraminifera, where water is clear and with slight currents

Identifying Features: Shell of this species is moderately small to moderately large, and moderately solid-to-solid. Last whorl is conical and occasionally narrowly conical or approaching conoid-cylindrical. Outline is slightly convex at adapical fourth, straight below. Shoulder is carinate. Spire is low and outline is variably concave to slightly sigmoid. Apex may project from an otherwise almost flat spire. Larval shell has 2 whorls. Later postnuclear whorls are carinate. Teleoconch sutural ramps are flat, slightly concave in later whorls, with pronounced axial threads. About 10–14 equidistant and evenly fine spiral grooves are seen on later ramps. Last whorl is with variably spaced weak spiral grooves on basal third, separating ribs near anterior end and ribbons above. Ground color is white. Last whorl is with a variable yellowish to dark brown pattern of reticulations and spiral bands. Larval shell is pale pink and darker pink posteriorly. Early teleoconch sutural ramps are pink to orange. Late sutural ramps are with yellowish to dark brown radial streaks and blotches coalescing with last whorl pattern and containing fine darker radial lines. Aperture is white, suffused with pale violet or pale brown. An adult shell varies between 29 mm and 71 mm in size. These snails are predatory and venomous. As they are capable of "stinging" humans, live ones should not be handled.

Toxin Type: m-conotoxin Nb3.1

Conus novaehollandiae **(Adams, A., 1854)**

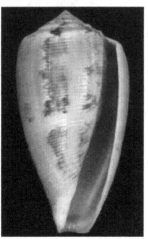

Apertural view

Common Name: Not designated

Geographical Distribution: Exmouth Gulf to Broome, W. Australia.

Habitat: Under stones; intertidal to shallow water.

Identifying Features: Like all species within the genus Conus, these snails are predatory and venomous. They are capable of "stinging" humans and therefore live ones should be handled carefully or not at all. This species largely resembles Conus anemone and nothing much is known about its biology

Toxin Type: A total of 161 proteins and protein isoforms have been identified from the venom glands of this species;

Three functionally active isoforms of peptidylprolyl cis-trans isomerise (PPlase) viz. PDI, PPI A, and PPI B (Helena et al., 2010); Arginine kinase (http://www.uniprot.org/uniprot/E3TMG6).

Conus nux **(Broderip, 1833)**

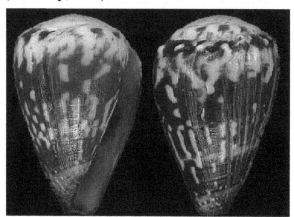

Apertural view   Abapertural view

Common Name: Nut cone

Geographical Distribution: From the Gulf of California to northern Peru, including the Galapagos, Clipperton, Revillagigedo, and Cocos Islands

Habitat: Shallow water species that occurs on sandy substrate and on rocks

Identifying Features: Shells of this species are very similar to those of C. sponsalis except for the larger size of the flammules on the body whorl. Adults grow to 30 mm but will typically be less than this. Like all species within the genus Conus, these snails are predatory and venomous. They are capable of "stinging" humans and therefore live ones should be handled carefully or not at all.

Toxin Type: T-superfamily, conotoxins nux5a and nux5b (Dovell, 2010);

A novel conopeptide sequence, member of the M-superfamily of conotoxins (Ramlakahan, 2002);

Four-loop conotoxin (http://www.uniprot.org/uniprot/B7SM24).

Conus ortneri **(Petuch, 1998) (=** *Conus* **(***Purpuriconus***)** *ortneri***)**

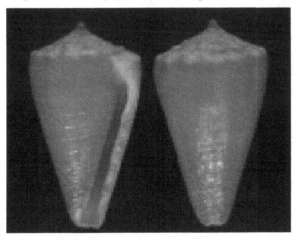

Abapertural view

Common Name: Not designated

Geographical Distribution: Bahamas: New Providence

Habitat: At depths of 6–7 m only

Identifying Features: Shell of this species is up to 25 mm, with high polish. Body whorl is characteristically sculptured with 6–8 evenly spaced, faint, shallowly depressed, punctate spiral grooves. Shoulder is highly rounded, subcarinated and is edged with 16 well-defined, rounded knobs per whorl. Shell color is uniformly deep orange-red or bright-orange, with paler knobs on spire. Protoconch and early whorls are deep cherry red and interior of aperture is rose-pink. Like all species within the genus Conus, these snails are predatory and venomous. They are capable of "stinging" humans and therefore live ones should be handled carefully or not at all.

Toxin Type: Alpha-conotoxin OmIA (http://www.ebi.ac.uk/interpro/signature/PS60014/proteins-matched;jsessionid=72887C4196F6CFF05EE34D33FCB8F5C0).

Conus parius **(Reeve, 1844)**

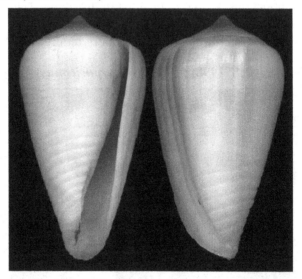

Apertural view   Abapertural view

Common Name: Parian cone

Geographical Distribution: Philippines and Indonesia to Papua New Guinea, Solomon Is. and Vanuatu

Habitat: In 2–20 m, on mud and muddy sand bottoms

Identifying Features: Shell of this species is moderately small to medium-sized, and moderately solid. Last whorl is ventricosely conical and outline is convex adapically, less so or straight below. Shoulder is subangulate or rounded. Spire is usually low and its outline is variably concave. Larval Shell is multispiral. Teleoconch sutural ramps are flat, with 0–1 increasing to 2–4 spiral grooves. Additional spiral striae are seen in latest whorls. Sculpture is usually weak on last ramp but with a distinct adaxial groove. Last whorl is with almost equally spaced axially striate spiral grooves on basal half. Color of the shell is gray or cream white, grading to light brown. Juveniles are dark brown. Some of these color tones often blend together both on last whorl and later sutural ramps. Shell may vary in size from 30 to 43 mm. Like all species within the genus Conus, these snails are predatory and venomous. They are capable of "stinging" humans and therefore live ones should be handled carefully or not at all.

Toxin Type: Conantokin-Pr1, -Pr2, and -Pr3 (http://www.uniprot.org/uniprot/P0C8E0);

Recombinant Conus parius Conantokin-Pr1 (http://www.biozol.de/products/mbs1003797-1-e/recombinant-conus-parius-conantokin-pr1-andere-spezies-auf-anfrage-1-mg.html);

αC-conotoxin PrXA (αC-PrXA) (A New Family of Nicotinic Acetylcholine Receptor Antagonists) (Jimenez et al., 2007);

M – superfamily, psi (ψ)-conotoxin (Lluisma et al., 2008)

Mu-conotoxin-like pr3b (http://www.cusabio.com/Recombinant-Protein/Recombinant-Conus-parius–Mu-conotoxin-like-pr3b-213175.html).

Conus parvatus (Walls, 1979)

Apertural view   **Abapertural view**

Common Name: Not designated

Geographical Distribution: From east South Africa (Natal) north to the Red Sea and continuing east, excluding India but including Sri Lanka, and to the west coast of Thailand

Habitat: Between the intertidal to depths of 5 m on reef flats, sand, rocks with algae, in rock crevices and dead coral

Identifying Features: The most striking characters of this species are the white band below the shoulder, and the spirally arranged reddish brown spots on the body whorl Tubercles of shoulder are sometimes obsolete. Ground color of shell is white to bluish-white or bluish-gray, usually lighter toward shoulder. Last whorl has spiral rows of varying numbers of reddish or brown dots and dashes from shoulder or near shoulder to base. Outer edges of late teleoconch sutural ramps have a spiral row of reddish-brown dots or lines. Base is dark bluish-violet. Aperture is pale violet to dark bluish-violet, sometimes with lighter bands below shoulder and centrally. Size of the shell varies from 9 to 24 mm. Like all species within the genus Conus, these snails are predatory and venomous. They are capable of "stinging" humans and therefore live ones should be handled carefully or not at all. This species has ornamental value as the shells are used in the necklace.

Toxin Type: Alpha-conotoxin (Balamurgan and Sivakumar, 2007; Balamurugan et al., 2008).

Conus pergrandis **(Iredale, T., 1937)**

Apertural view   Abapertural view

Common Name: Grand cone shell

Geographical Distribution: Taiwan; Mindanao, Basilan Island and Jolo Island in the Philippines; Papa New Guinea from West of Boigu Island to the West edge of Orangerie Bay; Inginoo on the North tip of Queensland to Deception Bay on the Australian coast; and New Caledonia, including Fayoué, Wé and La Roche

Habitat: At depths up to 400 m, most often 100–250 m.

Identifying Features: Adults of this species may grow to lengths of 90–185 mm. Like all species within the genus Conus, these snails are predatory and venomous. They are capable of "stinging" humans and therefore live ones should be handled carefully or not at all. Nothing much is known about its biology.

Toxin Type: α-superfamily, α-Conotoxin, PeIA (Adams and Belecki, 2013; McIntosh et al., 2005).

Conus pictus **Reeve, 1843**

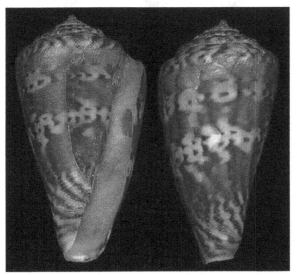

Apertural view   **Abapertural view**

Common Name: Not designated

Geographical Distribution: Endemic to South Africa; between Cape St Francis and the Northern Transkei region

Habitat: Shallow-water (20–60 m depth) and deep-water forms (50–100 m depth)

Identifying Features: Shell of this species has three broad pale scarlet bands in which the lower is ornamented with two articulated filets of brown and white. Spaces between the bands are variegated with brown (scarlet-brown) and the base and upper edge of the shell are obliquely streaked with the same color. Like all species within the genus Conus, these snails are predatory and venomous. They are capable of "stinging" humans and therefore live ones should be handled carefully or not at all.

Toxin Type: Conotoxin pc16a (Haegen et al., 2012).

Conus planorbis **(Born, 1778)**

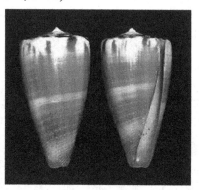

Abapertural view   **Apertural view**

Common Name: Planorbis cone

Geographical Distribution: Indo-West Pacific; across the central Indo-Pacific region from the Seychelles in the West, the coast of China to Japan in the North; across to French Polynesia in the East and Australia's North coast in the South

Habitat: Intertidal to about 60 m; on reef rock beneath dead coral, sand bottom with algae, and on coral and rubble.

Identifying Features: Shell of this species is medium-sized to moderately large and moderately solid-to-solid. Last whorl is conical or ventricosely conical and outline is convex at adapical fourth, almost straight below. Shoulder is angulate. Spire is low and is usually lower in form vitulinus. Outline is slightly concave, sigmoid or convex. Larval shell is of about 3 whorls. Teleoconch sutural ramps are flat and are often concave in late whorls, with 1 increasing to 5–7 spiral grooves. Spiral sculpture is occasionally weak on last 2 ramps. Last whorl is with variably raised and granulose spiral ribs are seen on basal third or fourth, sometimes weakly ribbed above. Ribs are variably spaced but are usually more closely set toward base. Ground color is white and is sometimes suffused with cream to tan on last whorl, but rarely so on sutural ramps. In typical form, last whorl is with a broad yellowish to dark brown spiral band on each side of center, sometimes blending with adjacent areas but usually leaving a ground color band at center and below shoulder; subshoulder band may be very narrow and interspersed with brown axial markings. Shell size varies from 40 to 82 mm. Like all species within the genus Conus, these snails are predatory and venomous. They are capable of "stinging" humans and therefore live ones should be handled carefully or not at all.

Toxin Type: J-superfamily, conotoxin pl14a (Elliger et al., 2011);

Conotoxin Pla_A (Pak, 2014);

P-superfamily, conotoxins- pl14a and pl9b;

O gene superfamily, conotoxins, pl6.1 (Imperial, 2007);

Pla_A06j, Pla_A06k, Pla_A06l, Pla_A06l, Pla_A04j, Pla_A04u, Pla_ B04f, Pla_B05g (Pak, 2014).

Conus princeps **(Linnaeus, 1758)** **(= *Ductoconus princeps*)**

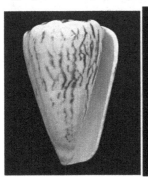

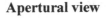

Apertural view   **Abapertural view**

Common Name: Prince cone

Geographical Distribution: From the Gulf of California south to northern Peru and the Galapagos Islands

Habitat: Intertidally in shallow water (1–30 m) on rocky reef habitat

Identifying Features: Prince Cone Shell, Conus princeps: The Prince Cone Shell is a striking and easily distinguishable shell with an orange-pink color with dark brown wavy axial stripes that extend up onto the spire. Spire is low and well coronated. Aperture is the same color as the outside of the shell but without stripes. It reaches a maximum length of 6.7 cm. It is a vermivorous species. Adults can reach a size of approximately 50 mm. Like all species within the genus Conus, these snails are predatory and venomous. They are capable of "stinging" humans and therefore live ones should be handled carefully or not at all. (http://www. mexfish.com/fish/princone/princone.html).

Toxin Type: A-conotoxins, pi1a-pi1d, pi1g, pi1h (These toxins are predicted to target diverse nicotinic acetylcholine receptor (nAChR) subtypes (Morales-González et al., 2015).

Conus purpurascens (Sowerby I & II, 1833)

Apertural view   Axial view   Abapertural view

Common Name: Purple cone

Geographical Distribution: from the Pacific coast of Mexico to northern Peru, including the Galapagos, Clipperton, Cocos, and Revillagigedo Islands.

Its feeds on fish and is considered the most venomous species in the eastern Pacific (Nybakken, 1970). Once adult it can reach a size of approximately 70 mm.

Habitat: Between low tide and 40 m under rocks on sandy mud substrate.

Identifying Features: Protoconch of this species is multispiral and shoulders are rounded, and surface of the shell has minute ridges. Periasrtracum is smooth and the operculum is small. Adult shell reaches a size of 70 mm. Its feeds on fish and is considered the most venomous species in the eastern Pacific. This snail is e capable of "stinging" humans and therefore live ones should be handled carefully or not at all.

Toxin Type: The venom of this species is the most powerful neurotoxin in the world and it has been reported to act within 0.002 second making it faster than the nerve conduction velocity. So the animal (victim) is down before it feels the prick of the dart. (http://jurassicpark.wikia.com/wiki/Conus_purpurascens);

Conotoxin p21a (Moller and Mari, 2011);

Conotoxin p5a (Dovell, 2010);

A-superfamily, conotoxin Pla and Plb;

Alpha-conotoxin PIVA (http://www.ebi.ac.uk/QuickGO/GProtein?ac=P55963);

mu-Conotoxin (http://www.t3db.ca/toxins/T3D2660);

Psi-conotoxin PIIIE (http://www.uniprot.org/uniprot/P56529);

psi-conotoxin pIIIe (Family: psi-conotoxin) – Superfamily: Conotoxins (http://scop.mrc-lmb.cam.ac.uk/scop/data/scop.b.ba.dc.b.e.b.html);

Psi-conotoxin PIIIF (http://www.cusabio.com/Recombinant-Protein/Recombinant-Conus-purpurascens–Psi-conotoxin-PIIIF-280777.html);

μ-Conotoxin PIIIA (Terlau and Olivera, 2004);

δ-PVIA (Terlau and Olivera, 2004);

κ-Conotoxin (http://www.t3db.ca/toxins/T3D2659);

κ-pVIIa (http://scop2.mrc-lmb.cam.ac.uk/6010232.html);

κ-conotoxin PVIIA (Terlau and Olivera, 2004);

κ-A-conotoxins PIVE and PIVF (kappa A-PIVE and kappa A-PIVF) (Teichert et al., 2007);

Contryphan-P and Leu-Contryphan-P (Gowd et al., 2005);

Conantokin-P.

Conus radiatus (Gmelin, 1791)

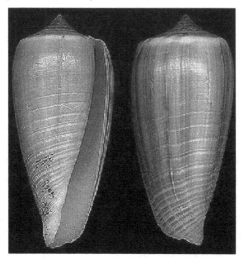

Apertural view   Abapertural view

Common Name: Scratched cone; Rayed cone

Geographical Distribution: Taiwan to Philippines, Papua New Guinea, Solomon Is. and Fiji.

Habitat: Common on sublittoral bottoms.

Identifying Features: Shell of this species is moderately small to large and moderately solid-to-solid. Last whorl is narrowly conoid-cylindrical, conoid-cylindrical or ventricosely conical. Outline is convex at adapical third, less so or straight below. Shoulder is subangulate or rounded. Spire is low and its outline is concave to straight. Larval shell is multispiral. Teleoconch sutural ramps are flat to slightly concave, with a pronounced subsutural ridge and 1 increasing to 6–8 spiral grooves. Additional spiral striae are seen on last ramp. Last whorl is with variably spaced and wide axially striate spiral grooves on basal third to half. Grooves are separating broad ribbons adapically and narrower ribbons and ribs near base. Last whorl is beige to dark brown, sometimes shaded with pale violet, often with a narrow white or gray spiral band at shoulder and less frequently with irregularly arranged white to bluish gray axial streaks. Larval shell is beige. Shell size varies from 30 to 109 mm. Like all species within the genus Conus, these snails are predatory and venomous. They are capable of "stinging" humans and therefore live ones should be handled carefully or not at all.

Toxin Type: Conotoxin κM-RIIIK (Terlau and Olivera, 2004);

Conophysin-R (Elliger et al., 2011; Terlau and Olivera, 2004);

I-superfamily, conotoxins r11a, r11b, r11c, r11d and r11e (Jimenez et al., 2003);

m-conotoxin R3.1;

Conantokin-C;

Des(Glyl)contryphan- R, Bromocontryphan-R and Contryphan- R (Gowd, et al., 2005).

Conus rattus **(Hwass in Bruguière, 1792)** *(= Conus (Rhizoconus) rattus)*

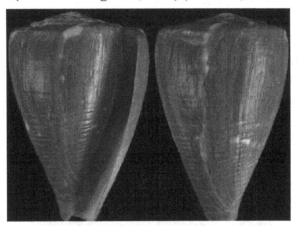

Apertural view   Abapertural view

Common Name: Rat cone

Geographical Distribution: Entire Indo-Pacific

Habitat: Fairly common on shallow reefs and tide pools

Identifying Features: Shell of this species is moderately small to moderately large, and moderately solid-to-solid. Last whorl is conical or ventricosely conical to broadly and sometimes broadly and ventricosely conical. Outline is variably convex at adapical third to three-fourths and left side is concave below. Shoulder is angulate. Spire is of low to moderate height and its outline is slightly convex to concave. Larval shell is of 3 or more whorl's with a maximum diameter 0.6–0.8 mm. Teleoconch sutural ramps are flat with 2–3 increasing to 3–6 spiral grooves. Last whorl is with variably prominent fine spiral ribs at base, gradually obsolete adapically. Ground color is bluish white to grayish blue. Last whorl is overlaid with

various shades of olive, brown or orangish brown, leaving a broad interrupted spiral ground-color band below shoulder and another obsolete to
broad one at center. Solid darker brown spiral lines may extend from base
to subshoulder area. On some portions of last whorl, brown spiral lines are
articulated with white dots producing a speckled appearance. Uniformly
dark brown shells intergrade with shells with numerous white dots and
blotches. Base is violet or dark brown. Shell size varies from 30 to 63 mm.
Like all species within the genus Conus, these snails are predatory and
venomous. They are capable of "stinging" humans and therefore live ones
should be handled carefully or not at all.

Toxin Type: Conotoxins Rt15a, Rt15b and Rt15c;
    XV conotoxin Rt15a (http://www.uniprot.org/uniprot/S4UJW7);
    XV conotoxin Rt15b;
    T superfamily conotoxin Rt5.1;
    Recombinant Alpha-conotoxin-like Rt20.2;
    Recombinant Alpha-conotoxin-like Rt20.1;
    w-Conotoxin Ra6.2;
    Alpha-conotoxin-like Rt20.2;
    m-conotoxin Ra3.1.

Conus regularis **(G. B. Sowerby II, 1833)** (= *Gradiconus regularis*)

Apertural view   **Abapertural view**

Common Name: Regular cone

Geographical Distribution: From the Pacific coast of Mexico to northern Peru

Habitat: Mud and sand between 10 and 60 m (Tenorio, 2013)

Identifying Features: It is a vermiverous species. Adults can reach a size of approximately 50 mm. Like all species within the genus Conus, these snails are predatory and venomous. They are capable of "stinging" humans and therefore live ones should be handled carefully or not at all. Nothing much is known about its biology.

Toxin Type: Omega-conotoxin RsXXVIA (http://www.uniprot.org/uniprot/P0DL31);
    Conotoxin RsXXIVA (Bernáldez et al., 2013).

Conus rolani (Röckel, 1986)

Apertural view   Abapertural view

Common Name: Cone snail

Geographical Distribution: Taiwan and the Philippines, Papua New Guinea and Makassar in Sulawesi, Indonesia

Habitat: At depths of between 60 and 400 m; sand and/or mud.

Identifying Features: Shell of this species is medium-sized to moderately large, and moderately solid-to-solid. Last whorl is ventricosely conical to pyriform and outline is convex at adapical half to two-thirds and straight to slightly concave (right side) or concave (left side) below. Shoulder is angulate, with 15–20 distinct to weak tubercles. Spire is of low to moderate height and its outline is concave or sigmoid. Larval shell is of 3 or more whorls with a maximum diameter about 0.9 mm. Teleoconch spire whorls are tuberculate. Teleoconch sutural ramps are slightly concave, with 0–2 increasing to 4–7 spiral grooves. Last whorl is with strong and is often granulose spiral ribs or ribbons. Ground color is white. Last whorl is with spiral rows of separate to fused brown dots, spots, and bars on ribs and ribbons, concentrated in 2 incomplete spiral bands, below shoulder and above center. Larval whorls and first 2–3 postnuclear sutural ramps are usually beige to brown. Following ramps are with sparse brown spots. Aperture is white. Shell size may vary from 46 to 70 mm. Like all species within the genus Conus, these snails are predatory and venomous. They are capable of "stinging" humans and therefore live ones should be handled carefully or not at all.

Toxin Type: Conantokin-Rl-A (Gowd et al., 2010);
    Conantokin-Rl-B;
    Conantokin-R1-C.

Conus sanguinolentus **(Quoy and Gaimard, 1834)**

    Apertural view   **Abapertural view**

Common Name: Bloodstained cone

Geographical Distribution: Throughout the entire Indo-Pacific; Hawaii (absent from the Central Indian Ocean)

Habitat: Shallow water at depths of 0.5–3 m; sand and rock boulders; under rocks on reef flats and at reef crest

Identifying Features: Shell of this species is moderately small to moderately large and moderately solid-to-solid. Last whorl is conical to slightly pyriform. Outline is variably convex at adapical third or half, straight to faintly concave below. Shoulder is angulate and is faintly to strongly tuberculate. Spire is of low to moderate height and its outline is straight to slightly concave. Postnuclear spire whorls are strongly tuberculate. Teleoconch sutural ramps are almost flat, with 1 increasing to 3–5 spiral grooves. Last whorl is with variably granulose spiral ribs on basal half, sometimes to subshoulder area. Last whorl is olive to orange brown, except for whitish granules on spiral ribs. Some specimens are with a slightly lighter central spiral band. Often evenly spaced brown spiral lines are seen from base to subshoulder area, abapically following granulated spiral ribs. Base and basal part of columella are purplish brown. Apex is yellowish white to bright orange. Late postnuclear sutural ramps are matching coloration of last whorl except for nearly white tubercles. Aperture is bright bluish violet behind a bright orange-brown marginal zone and whitish violet deeper within. Adults of this species are typically between 25–65 mm in length. This species feeds on polychaetes. Like all species within the genus Conus, these snails are predatory and venomous. They are capable of "stinging" humans and therefore live ones should be handled carefully or not at all.

Toxin Type: Conotoxins Sa0.1, Sa0.1 precursor, Sa1.1, Sa1.1 precursor, Sa1.10, Sa1.10 precursor, Sa1.11, Sa1.11, Sa1.11 precursor, Sa1.12, Sa1.12 precursor, Sa1.13, Sa1.13 precursor, Sa1.14, Sa1.14 precursor, Sa1.15, Sa1.15 precursor, Sa1.16, Sa1.16 precursor, Sa1.2, Sa1.2 precursor, Sa1.3, Sa1.3 precursor, Sa1.4, Sa1.4 precursor, Sa1.5, Sa1.5 precursor, Sa1.6, Sa1.6 precursor, Sa1.7, Sa1.7 precursor, Sa1.8, Sa1.8 precursor, Sa1.9, Sa1.9 precursor (http://www.conoserver.org/?page=list&table=protein& Organism_search%5B%5D=Conus%20 sanguinolentus);
I2-conotoxin (http://www.uniprot.org/uniprot/A0A061QL94);
O1-conotoxin (http://www.uniprot.org/uniprot/A0A061QHJ6);
Alpha-conotoxin(http://www.uniprot.org/uniprot/H9N3Y1);
Alpha-conotoxin-like (http://www.uniprot.org/uniprot/H9N3X1).

Conus sazanka (Shikama, T., 1970) (= *Kioconus sazanka*)

Abapertural view   **Apertural view**

Common Name: Sazanka's cone

Geographical Distribution: This species has three separate population groupings viz.: (1) East Africa from Somalia south to South Africa including Madagascar and Reunion Island; (2) Western Pacific from Hawaii to Japan and south to the Philippines including Indonesia; and (3) New Caledonia

Habitat: Between 50 and 200 m probably on sand and/or mud; on coral rubble

Identifying Features: Shell of this species is moderately small to medium-sized, and moderately light to moderately solid. Last whorl is conical and outline is almost straight to moderately convex with a constriction above base. Shoulder is subangulate to angulate and undulate to weakly tuberculate. Spire is of low to moderate height and its outline is concave. Larval shell of is about 4.0 whorls with a maximum diameter 0.90–0.95 mm. Postnuclear spire whorls are undulate. Teleoconch sutural ramps are flat, with 2 increasing to 4–5 spiral grooves. Last whorl is with a few faint spiral ribs at base. Color is reddish to brownish orange, occasionally yellow or light violet. Last whorl is usually with a lighter band around center, frequently interspersed with white flecks and adapically edged with brownish spots. There are white axial streaks below shoulder and across central band. Larval whorls are orangish pink to faint yellow. Teleoconch sutural ramps are variably maculated with darker yellow, pink or orange axial blotches. Aperture is translucent. Shell size varies from 25 to 42 mm. Like all species within the genus Conus, these snails are predatory and venomous. They are capable of "stinging" humans and therefore live ones should be handled carefully or not at all.

Toxin Type: α-conotoxin; this toxin has been reported to antagonize the rat neuronal nicotinic acetylcholine receptors (nAChRs) α3β2, α4β2, and α7 (Kauferstein, 2011).

Conus sponsalis **(Hwass in Bruguière, 1792)**

Abapertural view   Apertural view

Common Name: Sponsal cone, Marriage cone

Geographical Distribution: Entire Indo-Pacific; Aldabra, Chagos, Mascarene Basin, Mozambique, the Red Sea and the West Coast of South Africa

Habitat: Intertidal, and down to 100 m

Identifying Features: Shell of this species is small to moderately small, and moderately light to moderately solid. Last whorl is conical to broadly and ventricosely conical and is rarely slightly pyriform. Outline is convex at adapical half and usually straight below. In large specimens, aperture is often with a distinct spiral ridge at center. Shoulder is rounded to angulate, and is weakly to distinctly tuberculate. Spire is of low to moderate height, and its outline is concave to convex. Larval shell is of 4–5 whorls with a maximum diameter about 0.7 mm. Postnuclear spire whorls are finely tuberculate. Teleoconch sutural ramps are flat to slightly concave, with 1–4 spiral grooves, which are obsolete on late ramps. Last whorl is with fine, granulose spiral ribs on basal half. Ground color is white. Usual pattern of last whorl consists of reddish brown axial flames arranged in 2 spiral rows. Flames are often reduced in size or fusing into bands. Base and basal part of columella are purplish blue. Teleoconch sutural ramps are with reddish to blackish brown blotches between tubercles. Aperture is dark bluish violet deep within. Shell size varies from 15 to 34 mm. Like all species within the genus Conus, these snails are predatory and venomous. They are capable of "stinging" humans and therefore live ones should be handled carefully or not at all.

Toxin Type: A-superfamily, conotoxin Sp1.1;

Four – loop conotoxin (http://www.uniprot.org/uniprot/B7SLZ2).

Conus spurius (Gmelin, 1791) (= *Conus bahamensis*)

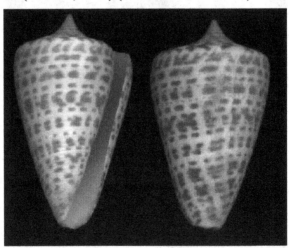

Apertural view   Abapertural view

Common Name: Alphabet cone

Geographical Distribution: Gulf of Mexico.

Habitat: At depths of 0–64 m; coral reefs or shallow ocean

Identifying Features: This species has a concave spire, with the first 4–6 whorls slightly tuberculate, the last whorl pyriform and with a sutural ramp that has a broad spiral groove. Color of the shell is yellowish-white, usually with a broad whiter band at the center of the last whorl and near the shoulder. Maximum-recorded shell length is 80 mm. This is a vermicorous species. Like all species within the genus Conus, these snails are predatory and venomous. They are capable of "stinging" humans and therefore live ones should be handled carefully or not at all.

Toxin Type: Conorfamide (Terlau and Olivera, 2004);
    T-1 superfamily toxins (Zamora-Bustillos et al., 2009);
    Conotoxin sr5a (Dovell, 2010);
    I-superfamily, conotoxin sr11a (Aguilar et al., 2007);
    I-superfamily, conotoxins sr11.1 to sr11.10 (http://www.ebi.ac.uk/interpro/entry/IPR020242/proteins-matched?start=40);
    A-superfamily, conotoxins SrIA and SrIB.

Conus stercusmuscarum **(Linnaeus, 1758)**

Apertural view   Abapertural view

Common Name: Fly specked cone

Geographical Distribution: Japan to Marshall Is. and to Indonesia, Papua New Guinea and Solomon Is.; probably also Fiji.

Habitat: Intertidal and uppermost subtidal; in sand and beneath corals.

Identifying Features: Shell of this species is medium-sized to moderately large and moderately solid-to-solid. Last whorl is usually conoid-cylindrical to nearly cylindrical or slightly ovate and outline is slightly convex. Siphonal fasciole is prominent. Shoulder is sharply angulate. Spire is of low to moderate height and its outline is straight to slightly convex. Larval shell is multispiral with maximum diameter about 0.6 mm. First 3 postnuclear whorls are weakly tuberculate. Teleoconch sutural ramps are flat in early whorls and concave in late whorls, with 2 increasing to 3–4 spiral grooves. Last whorl is with regularly spaced, broad spiral ribs, distinct basally but obsolete adapically. Ground color is white to pale gray. Last whorl is with spiral rows of irregularly alternating blackish brown dots and white dashes or bars. Dark dots are found clustered into patches forming 2 interrupted spiral bands of both sides of center. Larval whorls are pink. Teleoconch sutural ramps are with blackish brown markings along inner and outer margins, partially connected across the ramps. Aperture is orange deep within. It probably preys on fishes. Shells of this range in size from 40 to 64 mm. These snails are predatory and venomous. They are capable of "stinging" humans and therefore live ones should be handled carefully or not at all.

Toxin Type: Mu-conotoxin SmIIIA (http://www.uniprot.org/uniprot/ P60207);

Delta-conotoxin-like SmVIA (Delta-SmVIA) (www.cusabio.com.-2007–2015);

Conotoxin superfamilies A, O and M; contryphan;

μ-conotoxin SmIIIA (Terlau and Olivera, 2004);

Contryphan-Sm (Gowd et al., 2005);

Alpha-conotoxin-like Sm1.1 (http://www.ebi.ac.uk/interpro/signature/ PS60014/proteins-matched;jsessionid=72887C4196F6CFF05EE34D33F CB8F5C0).

Conus striolatus (Kiener, 1845)

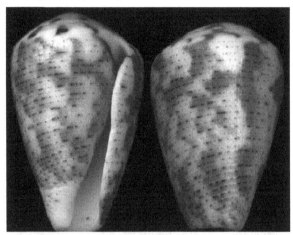

Apertural view   Abapertural view

Common Name: Broad-streaked cone.

Geographical Distribution: Pacific Ocean from Thailand to Micronesia and from Taiwan to Queensland, Australia.

Habitat: Inter-tidal and shallow subtidal to depths of 20 m where it lives in muddy sand

Identifying Features: Shell of this species is small and solid. Body whorl is ventricosely conical. Shoulder is rounded. Spire is of moderate height and its outline is straight. Body whorl is with spiral ribs at base. Basal spiral ribs on both sides are finely granulated, and are conspicuously prominent on the ventral side. Ground color is pale gray. Body whorl is with brown axial blotches, fusing into interrupted spiral bands on each side of center. Spiral rows of alternating brown to black and white dots and dashes extend from base to shoulder. Aperture is white. Size of an adult shell varies between 20 mm and 46 mm. These snails are predatory and venomous. As they are capable of "stinging" humans, live ones should be handled.

Toxin Type: Mu-conotoxin SxIIIA;
    Mu-conotoxin-like SxIIIB;
    Kappa-conotoxin-like Sx11.2 (http://www.ebi.ac.uk/interpro/entry/IPR020242/proteins-matched?start=40).

Conus sulcatus (Hwass, 1792)

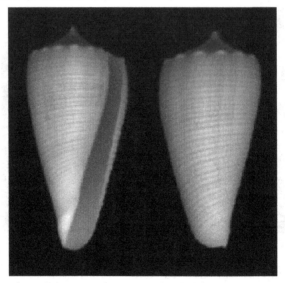

Apertural view   **Abapertural view**

Common Name: Sulcate cone

Geographical Distribution: From India and Sri Lanka, across the whole of South-east Asia, north to Japan, south to northern Australia and east to Fiji.

Habitat: At depths of 20–240 m where it lives on sand

Identifying Features: Shell of this species is medium-sized to large and moderately solid-to-solid. Last whorl is conical to ventricosely conical, broadest in form samiae. Outline is almost straight and left side sometimes is concave near base. Shoulder is angulate and strongly tuberculate to weakly undulate with about 10–14 broad tubercles or bulges. Spire is of low to moderate height and its outline is concave to almost straight. Larval shell is of 3 or more whorls with a maximum diameter 0.8–1 mm. Teleoconch spire whorls are tuberculate to undulate. Teleoconch sutural ramps are concave, with 1–2 increasing to 5–12 spiral grooves. Ground color is white. Last whorl is variably suffused or streaked with yellowish to dark brown. Almost uniformly brown shells intergrade with largely white shells. Shoulder edge is usually white. Larval whorls and adjacent postnuclear sutural ramps are immaculate, following sutural ramps maculated with brown radial markings, ranging from mainly brown to mainly white. Aperture is white. Shell size varies from 48 to 89 mm. These snails are predatory and venomous. As they are capable of "stinging" humans, live ones should be handled.

Toxin Type: A-superfamily conotoxin Su1.1 to Su1.8;
    Conantokin-Br (ConBr);
    Conantokin-B1 (http://www.conoserver.org/?page=list&table=protein &Organism_search%5B%5D=Conus%20 sulcatus).

Conus terebra **(Born, 1778)**

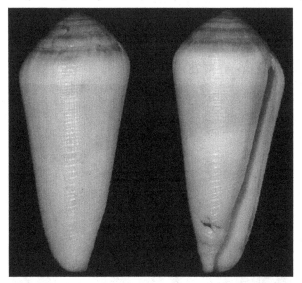

Abapertural view   Apertural view

Common Name: Terebra cone

Geographical Distribution: Throughout the tropical Indo-Pacific

Habitat: At depths of 0.5–40 m on coral reefs, coral rubble and in fine sand with algae

Identifying Features: Shell of this species is moderately small to large with low gloss. Body whorl is conical to narrowly conical and outline is convex at shoulder, straight below, sometimes concave centrally. Shoulder is rounded to roundly angled, not distinct from spire. Spire is of moderate height and its outline is convex. Body whorl is with variably spaced and variably fine spiral ribs from base to shoulder. Spiral ribs are generally closer near shoulder. Aperture is narrow and is slightly wider anteriorly. Outer lip is straight. Ground color is white to pale cream. Body whorl is with a broad spiral band on each side of the center varying from light gray to light brown. Base is tinged with violet in adult specimens. Aperture is white, in adult pale or dull violet. Adults of the species will grow to approx. 110 mm although they will typically be less than this. This is a vermiverous species, which eats polychaetes. These snails are predatory and venomous. As they are capable of "stinging" humans, live ones should be handled.

Toxin Type: O3-superfamily, conotoxin Tr7.5;
    O2-superfamily, conotoxin Tr7.2;
    T-superfamily, conotoxin Tr5.2;

Conus tessulatus **(Born, 1778)**

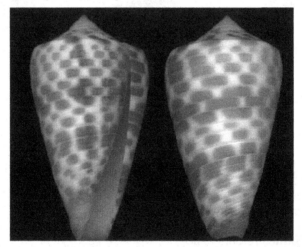

Apertural view   **Abapertural view**

Common Name: Tessellate cone

Geographical Distribution: From the east coast of Africa across the Indian Ocean and Pacific Ocean to the west coast of Central America from Western Mexico to Costa Rica.

Habitat: Intertidal and subtidal areas of coast of depths of up to around 40 m; coral reefs in bays with a fine to coarse sand; muddy sand and on gravel in sheltered flats.

Identifying Features: Shell of this species is medium sized to moderately large and solid to moderately heavy. Body whorl is conical to broadly conical. Outline is convex near shoulder and straight below. Shoulder is broad and subangulate to angulate. Spire is low to moderate height and sharply pointed and its outline slightly concave. Aperture is moderately narrow and with almost uniform width. Outer lip is sharp and straight. Body whorl is with variously spaced, weak or incised or often punctuates spiral grooves on apical third. Ground color is white. Body whorl is with spiral rows of mostly bright orange rectangular spots or bars, often alternating with white markings. These color markings usually fuse into spiral bands in each side of the center. Shoulder and spire are with radial markings matching bars on last whorl in size and color. Base is bluish-white. Aperture is white, usually with pink tones. This species grows to between 32 and 80 mm. These snails are predatory and venomous. As they are capable of "stinging" humans, live ones should be handled.

Toxin Type: Conotoxin TsMMSK-011 (http://www.uniprot.org/uniprot/Q9BPI6);

Conotoxin TsMMSK-021(http://www.cusabio.com/Recombinant-Protein/Recombinant-Conus-tessulatus–Conotoxin-TsMMSK-021–909909.html);

Conotoxin TsMLKM-012 (http://www.cusabio.com/Recombinant-Protein/Recombinant-Conus-tessulatus–Conotoxin-TsMLKM-012–908660.html);

Alpha-conotoxin-like ts14a (http://www.uniprot.org/uniprot/P86362);

A-superfamily, conotoxin ts14a;

T-superfamily, conotoxins Ts011, Ts03, Ts5.1 to Ts5.4 and Ts5.7;

Superfamily? Conotoxins Ts14a, Ts5.5, TsMEKL-02 (partial) and TSMEKL-03;

M-superfamily, conotoxins Ts3-IP07, Ts3-IP10, Ts3-SGNO1, Ts3-Y01, Ts3.1 to Ts3.7 and TsMMSK-021;

01-superfamily, conotoxin Ts6.1;

02-superfamily, conotoxins Ts6.2 and Ts6.3;

03-superfamily, conotoxins Ts6.4 to Ts6.7 (http://www.conoserver.org/?page=list&table=protein&Organism_Clade%5B%5D=XVI&Type%5B%5D=Wild+type);

δ(delta)conotoxinTsVIA(Amanetal.10.1073/pnas.1424435112);Tessula toxin (vasoactive protein);

This toxic protein causes a marked contraction of the rabbit isolated aorta (Kobayashi et al., 1983).

Conus tinianus **(Hwass in Bruguière, 1792) (=** *Ketyconus tinianus***)**

Apertural view   Axial view   Abapertural view

Common Name: Variable cone

Geographical Distribution: Endemic to east coast of South Africa; from Tongaat to Cape Agulhas

Habitat: Sandy areas on reefs under large rocks; associated with lace coral; from intertidal shallow waters to 40 m

Identifying Features: This species grows to approx. 60 mm. Like all species within the genus Conus, these snails are predatory and venomous. They are capable of "stinging" humans and therefore live ones should be handled carefully or not at all. Nothing much is known about its biology.

Toxin Type: Alpha-conotoxin (http://www.uniprot.org/uniprot/P0CI05).

Conus varius (Linnaeus, 1758)

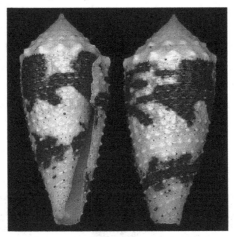

Apertural view   Abapertural view

Common Name: Freckled cone

Geographical Distribution: Across the Indian Ocean from the East coast of Africa to the Indo-Pacific region, including northern Australia.

Habitat: Coral reefs, rock substrate, and mangrove areas; hard coral, rubble, and turf algae; at depths from intertidal to about 240 m

Identifying Features: Typical size for shells of this species is between 35–61 mm in length. Like all species within the genus Conus, these snails are predatory and venomous. They are capable of "stinging" humans and therefore live ones should be handled carefully or not at all. Nothing much is known about its biology.

Toxin Type: T superfamily conotoxin Vr5.2 (http://www.uniprot.org/uniprot/S4UKK8);

A-superfamily, conotoxins Vr1.1, Vr1.1 precursor, Vr1.2 and Vr1.2 precursor;

O2-superfamily, conotoxins Vr15a, Vr15a precursor, Vr15b, Vr15b precursor;

M superfamily, conotoxins Vr3–3-VP01, Vr3–3-VP01 precursor, Vr3-D01, Vr3-D01 precursor, Vr3-IP01, Vr3-IP01 precursor, Vr3-IP03, Vr3-IP03 precursor, Vr3-IP08, Vr3-IP08 precursor, Vr3-L01, Vr3-L01 precursor,

Vr3-NPP01, Vr3-NPP01 precursor, Vr3-Q01, Vr3-Q01 precursor, Vr3-SP01, Vr3-SP01 precursor, Vr3-SP02, Vr3-SP02 precursor, Vr3-SP04, Vr3-SP04 precursor, Vr3-T05, Vr3-T05 precursor, Vr3-TYN01, Vr3-TYN01 precursor, Vr3-VP08, Vr3-VP08 precursor, Vr3-WP04, Vr3-WP04 precursor, Vr3-Y02, Vr3-Y02 precursor, VrD01 and Vr-DO1 precursor;

T-superfamily, conotoxins Vr5.1, Vr5.1 precursor, Vr5.2, Vr-5.2 precursor, Vr5.3, Vr5.3 precursor Vr5.4 and Vr-5.4 precursor; (http://www.conoserver. org/?page=list&table=protein&Organism_search%5B%5D=Conus%20 varius).

Conus ventricosus **(Gmelin, 1791)**

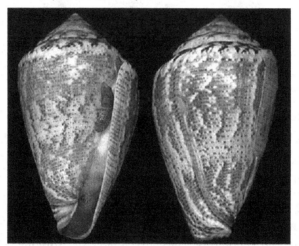

Apertural view   **Abapertural view**

Common Name: Mediterranean cone

Geographical Distribution: Entire coast of the Mediterranean including its Atlantic approaches from the Algarve, Portugal to Huelva and Cadiz, Spain, and from Rabat, Morocco towards Gibraltar

Habitat: Shallow waters in many differing habitats including mud, sand, seagrass and rocks between 0.5 and 5 m depth

Identifying Features: Shell of this species is turgid in shape with convex sides. Protoconch is paucispiral. Whorl tops are ornamented with cords that reach the middle spire whorls and often persist. Anal notch is shallow to moderate in depth. Periostracum is smooth and thin, and the operculum is of moderate size. Adults of the species typically grow to 45 mm in length. These snails are predatory and venomous. They are capable of "stinging" humans and therefore live ones should be handled carefully or not at all.

Toxin Type: Contryphan-Vn (Gowd et al., 2005);
Omega-conotoxin-like Vn2 (http://www.uniprot.org/uniprot/P83301);
Alpha-conotoxin-like (http://www.uniprot.org/uniprot/P0C8 V4);
Conotoxin VnMRCL-04;
Conotoxin VnMEKL-0222 (http://www.uniprot.org/uniprot/Q9BPC9).

Conus vexillum **(Gmelin, 1791)**

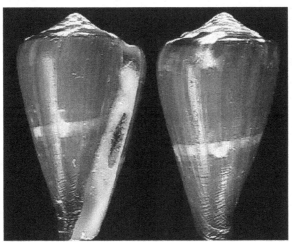

Apertural view   Abapertural view

Common Name: Vexillum cone, flag cone

Geographical Distribution: Across the entire Indo-Pacific from Natal to Hawaii and French Polynesia and Japan to Australia.

Habitat: Juveniles are present on intertidal benches whereas adults occur along subtidal reefs to 50–70 m; shallow water, lagoon pinnacles, sand, sand with gravel, among weed or rocks and under dead coral.

Identifying Features: Shell of this species is medium sized to large and solid to heavy. Body whorl is conical to broadly conical and outline is convex adapically and straight below. Shoulder is angulate to rounded. Spire is of low to moderate height and its outline is straight to slightly convex and weak spiral ribs are on base. Aperture is uniformly wide and interior is white. Outer lip is thin sharp and straight. Ground color is white. Body whorl is brown except for variably broad white spiral bands at center and shoulder. Spire is white, flecked with brown. Base is dark-brown. Adults of the species typically grow between 65 and 183 mm. This species feeds on eunicid polychaetes. These snails are predatory and venomous. They are capable of "stinging" humans and therefore live ones should be handled carefully or not at all.

Toxin Type: αD-conotoxins VxXIIA and VxXIIB (Grosso et al., 2014);
αD-conotoxin VxXXA (Loughnan et al., 2009);
αD-conotoxin VxXXC;
O-superfamily, conotoxins vx6a and vx6b (Jiang, et al., 2006);
Kappa-conotoxin-like 1, Kappa-conotoxin-like 2; (http://www.ebi. ac.uk/interpro/entry/IPR020242/proteins-matched?start=40).

Conus victoriae (**Reeve, 1843**)

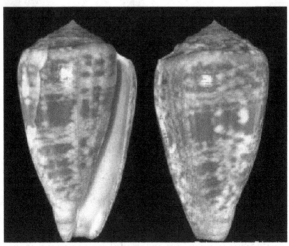

Apertural view   **Abapertural view**

Common Name: Queen Victoria cone

Geographical Distribution: Western Australia from Broome north to the mouth of the Victoria River, Northern Territory

Habitat: Intertidal to about 10 m; on mud and sand bottoms of coral reef, beneath and among rocks and in rock pools, exposed or buried in sand

Identifying Features: Shell of this species is medium-sized to moderately large and moderately solid-to-solid. Last whorl is ventricosely conical, ovate or conoid-cylindrical. Outline is convex and is often less so to nearly straight below adapical third. Left side is concave to nearly straight at basal third. Shoulder is angulate to sometimes subangulate. Spire is of low to moderate height and its outline is usually concave to straight. Larval shell is of 1.75–2 whorls with a maximum diameter 1–1.1 mm. First 3.5–6.5 postnuclear whorls are tuberculate. Teleoconch sutural ramps are flat to slightly concave, with 1 increasing to 6–12 variably fine, weak to obsolete spiral grooves. Last whorl is with variably spaced, distinct to obsolete spiral ribs basally. Ground color is white and is often suffused with light blue and/or overlaid with yellowish or orangish brown. Color pattern is extremely variable. Last whorl in typical form is with a network of fine light to blackish brown lines edging very small to medium-sized tents. Yellowish or orangish to blackish brown blotches are arranged in 2–3 or more interrupted to continuous spiral bands and interspersed with coarse darker axial lines. Shells with typical pattern are intergrade with shells with an axial pattern of bands and streaks to closely spaced fine wavy axial lines. Pattern varies from completely white shells to shells heavily patterned on a brown or blue ground. It is a mollusc-eating cone (mollusiovore). It measures 53 × 27 mm in shell length. Like all species within the genus Conus, these snails are predatory and venomous. They are capable of "stinging" humans and therefore live ones should be handled carefully or not at all.

Toxin Type: Toxins of I1-superfamily, I2-superfamily and I4-superfamily; Toxins of A-superfamily, M-superfamily, O1 superfamily, O2 superfamily, O3-superfamily, Contryphan superfamily, P-superfamily, S-superfamily, T-superfamily, Conantokins (B-superfamily), B2-superfamily, E- and F-superfamilies, U-superfamily (Robinson et al., 2014);

Con-ikot-ikot precursor sequence;

A- superfamily, conotoxin Vc1a;

Conotoxin Vc1.1 (ACV1) (http://grimwade.biochem.unimelb.edu.au/cone/new2008.html);

Conotoxin vc5c (Dovell, 2010);

Alpha-conotoxin Vc1.2;

Conotoxin I2_Vc11.10 (prepropeptide);

Conotoxin I2_Vc11.9 (prepropeptide);

Conotoxin I2_Vc11.7 (prepropeptide) (http://www.ebi.ac.uk/interpro/entry/IPR020242/proteins-matched?start=40).

Conus villepinii (Fischer and Bernardi, 1857)

Apertural view   Abapertural view

Common Name: Villepin's cone

Geographical Distribution: Gulf of Mexico

Habitat: At depths of 25–475 m.

Identifying Features: It has a broadly conical, moderately elevated spire, which is concave in profile and with the suture running sufficiently below the rounded peripheral shoulders of each preceding whorl. Body whorl just below periphery may be slightly convex. Body whole profile is straight to slightly concave along the lower portion viewed from the apertural side. Entire body is sculptured with spiral grooves usually weaker posteriorly being more pronounced and angled anteriorly. Aperture is narrow and slightly flared at anterior end. Outer lip is thin and interior is white. Maximum-recorded shell length is 93 mm and it is a worm-hunting snail. Like all species within the genus Conus, these snails are predatory and venomous. They are capable of "stinging" humans and therefore live ones should be handled carefully or not at all.

Toxin Type: ConoCAP (http://www.uniprot.org/uniprot/E3PQQ8);
Conotoxin vil14a (Möller et al., 2005);
F14 conotoxin vil14a (Dovell, 2010; Elliger et al., 2011; Moller et al., 2005);
CCAP-vil (C-terminally amidated decapeptide) (Miloslavina et al., 2010).

Conus virgo (Linnaeus, 1758)

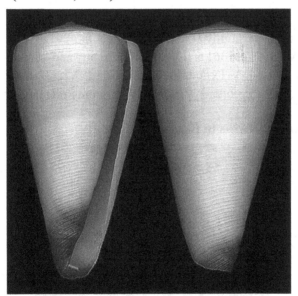

Apertural view   Abapertural view

Common Name: Virgin cone

Geographical Distribution: Entire Indo-Pacific apart from the Hawaiian Islands

Habitat: In 0.5–15 m; in sand and rubble on reef flats, sometimes among weed and beneath dead coral rocks

Identifying Features: Shell of this species is moderately large to large and solid to heavy. Last whorl is conical. Outline is slightly convex at adapical fourth and straight below. Shoulder is angulate. Spire is low and its outline is slightly concave to slightly convex. Teleoconch sutural ramps are almost flat to slightly concave. Late ramps are with 3 increasing to 5–6 spiral grooves, either paralleled by additional striae or replaced by numerous striae in latest whorls. Last whorl is with weak to obsolete spiral ribs near base. Widely spaced fine ribs and wrinkled threads may extend to center or beyond. Color is white to yellow or orange and occasionally with darker orange collabral lines marking growth cessations. Base is dark blue-violet. Larval whorls are bright purple (eroded in adults). Aperture is white, blue-violet at base. Size of the shell varies from 55 to 151 mm. It feeds on terebellid and other polychaetes. These snails are predatory and venomous. They are capable of "stinging" humans and therefore live ones should be handled carefully or not at all.

Toxin Type: T-superfamily conotoxins (Liu et al., 2012);

C.virgo Vi1359, Vi 1361 and Vi805 (Ramasamy and Manikandan, 2011);

Kappa-conotoxin ViTx and conotoxins Vi11.2, Vi11.3 and Vi11.5 (http://www.ebi.ac.uk/interpro/entry/IPR020242/proteins-matched?start=40);

Conus vitulinus (Hwass in Bruguière, 1792)

Apertural view   Abapertural view

Common Name: Necklace cone, calf cone

Geographical Distribution: North-east Indian Ocean along India and Sri Lanka to West. Thailand

Habitat: Reef; On the beach reef

Identifying Features: Shell of this species is medium to large, moderately heavy, and conical with high gloss. Sides are nearly straight. Body whorl is conical and outline is variably convex over adapical fourth to third and straight below. Body whorl is with a few weak spiral ridges above the base and sometimes-conspicuous axial threads. Shoulder is broad, carinate to angulate and concave above. Spire is of low to moderate height. Ground color is white or cream. Body whorl is suffused or spirally banded with pale orange or pink. Spiral rows are of brown dots, dashes and variously shaped spots extend from base to shoulder but vary in number and arrangement, often concentrated at both sides of the center. Sometimes dark markings fuse into axial flames or blotches. Base is pale orange or brown. Aperture is white. Size of an adult shell varies between 45 mm and 95 mm. It is a worm eater. Like all species within the genus Conus, these snails are predatory and venomous. They are capable of "stinging" humans and therefore live ones should be handled carefully or not at all.

Toxin Type: Conotoxins Vi11.3 and Vi11.7 (http://www.ebi.ac.uk/interpro/entry/IPR020242/proteins-matched?start=40);
    Conotoxins Vt15a and Vt15b.

Conus wittigi **(Walls, 1977)**

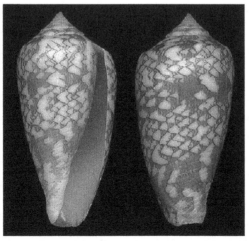

Apertural view		**Abapertural view**

Common Name: Not designated

Geographical Distribution: Endemic around the Lesser Sunda Islands, Flores and the North of Timor

Habitat: Shallow water species occurring at depths between 3 and 10 m; on fine sand near coral

Identifying Features: Shell of this species is moderately small to medium-sized, and is generally moderately light. Last whorl is ventricosely conical to conoid-cylindrical. Outline is slightly convex at adapical two-thirds and straight below. Left side may be sigmoid. Aperture is wider at base than near shoulder. Shoulder is angulate. Spire is of low to moderate height and its outline is concave. Larval shell is of 2.25–2.5 whorls with a maximum diameter 0.8–0.9 mm. Teleoconch sutural ramps are flat, with 1 increasing to 3–4 major spiral grooves. Basal third to half of last whorl is with wide spiral grooves at base and narrow grooves above. Ribbons are seen between grade to ribs at anterior end. Ground color is white. Last whorl is with yellow to red brown reticulated lines, edging variably sized white tents and blotches. Pattern fuses into very interrupted to solid spiral bands of varying width above and below center. Larval whorls are beige. Postnuclear sutural ramps are with orange to brown radial lines and blotches. Aperture is white. Adults of this species typically grow to between 26–42 mm. Like all species within the genus Conus, these snails are predatory and venomous. They are capable of "stinging" humans and therefore live ones should be handled carefully or not at all.

Toxin Type: m-conotoxin Wi3.1 (LookChem- http://www.lookchem.com/cas/394743–01–2_394744–00–4.html).

Conus zeylandicus **(Gmelin, 1791)**

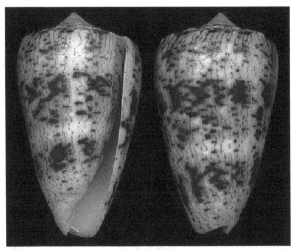

Apertural view   Abapertural view

Common Name: Obese cone

Geographical Distribution: Throughout the Indian Ocean, excluding the Red Sea

Habitat: Mud, sand, and rubble substrate to depths of 35 m

Identifying Features: Shell of this species is medium-sized to moderately large and solid to moderately heavy. Last whorl is conical, conoid-cylindrical or broadly and ventricosely conical. Outline is convex above base and below shoulder, almost straight in between. Shoulder is rounded and irregularly undulate. Spire is of low to moderate height and its outline is straight to deeply concave. Larval shell is of about 2.75 whorls with a maximum diameter 0.8–1 mm. Postnuclear spire whorls are tuberculate, less so toward shoulder. Teleoconch sutural ramps are flat. Concave in late whorls, with 1 increasing to 3–4 spiral grooves. Last whorl is with variably broad spiral ribs at base. Ground color is white and is suffused with cream, pink or violet. Last whorl is with gray to violet clouds. Dark brown elements are usually aligned in 2 incomplete spiral bands, on both sides of center, often forming a weak additional band below shoulder. Shell size varies from 45 to 75 snails are predatory and venomous. They are capable of "stinging" humans and therefore live ones should be handled carefully or not at all.

Toxin Type: The lethality of the crude venom (LD50–60 mg/kg via i.p.) of this species in mice was associated with increased heart rate and strong muscular hind limb paralysis, skeletal muscle paralysis, dyspnea, loss of spontaneous activity followed by respiratory failure. Further, liver tissues were disrupted with hemorrhagic necrosis and the lung showed the pathogenic changes of diffused inflammation of the parenchyma and obliteration of the alveolar space. In brain, edema was observed throughout the parenchyma and the kidney showed the tubules with cloudy swelling of the lining cells and the parenchyma inflammation and few inflammatory cells infiltration (Annadurai et al., 2007).

Cylinder abbas **(Hwass in Bruguière, 1792)** (= *Conus abbas*)

Apertural view   **Abapertural view**

Common Name: Abbas Cone

Geographical Distribution: East Africa, Ceylon, Philippines, New Caledonia

Habitat: Shallow water to a maximum depth of about 50 m on coral reefs, often beneath coral boulders

Identifying Features: Shell of this species is white, very finely reticulated with narrow orange-brown lines, with a broad central and often narrower upper and lower bands of darker color bearing occasional longitudinal chocolate stripes. The height of the shell is from 38 mm to 64 mm. Snails of this species are predatory and venomous. They are capable of "stinging" humans and therefore live ones should not be handled.

Toxin Type: Conopeptides; venom type not reported.

Conus aculeiformis **(Reeve, 1844) (=** *Conasprella aculeiformis***)**

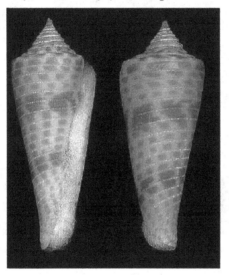

Apertural view   Abapertural view

Common Name: Spindle cone

Geographical Distribution: Red Sea, S. Persian Gulf, S. E. India to Andaman Is., and S. Indonesia

Habitat: Habitat and ecology of this species are unknown, as locality data are uncertain. Reports show depth recorded at 50–100 m

Identifying Features: This species has a narrow shell, which has an elevated spire. It is encircled with equidistant punctate grooves, and flat interspaces. Shoulder is angulate to carinate, with a deep exhalent notch. Last whorl is narrowly conical and outline is almost straight. Aperture pale brown, sometimes darker deep within. Color of the shell is white, with light chestnut spots and two interrupted broad bands of chestnut clouding. Shell attains a maximum height of 4 cm. Like all species within the genus

Conasprella, these snails are predatory and venomous. They are capable of "stinging" humans and therefore live ones should not be handled.

Toxin Type: Conopeptides; venom type not reported.

Conus acutangulus (**Lamarck, 1810**) (*= Conus (Turriconus) acutangulus*)

Abapertural view and vice-versa   Abapertural view

Common Name: Sharp-angled cone

Geographical Distribution: Throughout the Indo-Pacific, ranging from West Africa to Hawaii and French Polynesia

Habitat: Sublittoral (5–100 m); occurs on sand, coral rubble, muddy sand and seaweed.

Identifying Features: Shell of this species is medium-sized and moderately solid. Body whorl is conical and sides are nearly straight with broad raised ribs of fairly uniform width separated by narrow deep grooves. Ground color is white. Body whorl is light brown with scattered blotches of ground color at shoulder and center. Shoulder is carinate. Spire is high and is marked by distantly spaced brown spots and streaks. Aperture is narrow, and interior is pale brown. Outer lip is thin and fragile. Adults can grow up to 38 mm in length, although they will typically be smaller than this. These snails are predatory and venomous. As they are capable of "stinging" humans, live ones should not be handled carefully or not at all

Toxin Type: Conopeptides; venom type not reported.

Conus advertex **(Garrard, 1961) (=** *Plicaustraconus advertex, Conus angasi f. advertex***)**

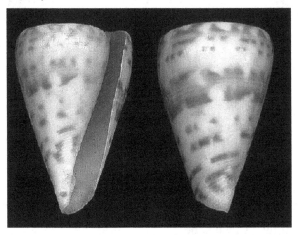

Apertural view   Abapertural view

Common Name: Not designated

Geographical Distribution: Australian and South African regions.

Habitat: This species is found between 130 m and 250 m in sand with coarse shell rubble.

Identifying Features: Shell of this species is obconic in shape with broad shoulders. Protoconch is paucispiral. Whorl tops have cords, and there is a well-developed dentiform plait. Anal notch is shallow to relatively deep, and an anterior notch is absent. Periastracum is smooth and the operculum is large. Anterior section of the radular tooth is shorter than the posterior section. Barb is short. There are one or two rows of serrations. Once mature, it can reach a size ranging from 30 to 46 cone snails are presumed vermivorous, preying on polychaete worms. These snails are predatory and venomous. They are capable of "stinging" humans and therefore live ones should not be handled.

Toxin Type: Conopeptides; venom type not reported.

Conus africanus **(Kiener, 1845)** (= *Varioconus africanus*)

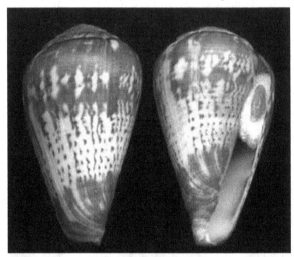

Abapertural view and vice-versa   Abapertural view

Common Name: African cone

Geographical Distribution: West Africa; Angola

Habitat: There is little information on habitat in the literature

Identifying Features: Shell of this species is small, moderately high and moderately solid. Last whorl is broadly and ventricosely conical to ovate and shoulder is rounded. Outline of the shell is convex at adapical third, near base. Spire is low to moderately high and straight to slightly convex. Last whorl is smooth and dull, with about 10 spiral grooves at base. Ground color of the shell is white, with dark brown blotches and streaks turning into bars and dashes. Aperture white. This species grows to a maximum size of 35 mm. Like all species within the genus Conus, these snails are predatory and venomous. They are capable of "stinging" humans and therefore live ones should not be handled.

Toxin Type: Conopeptides; venom type not reported.

Kioconus alconnelli **(da Motta, 1986) (=** *Conus alconnelli***)**

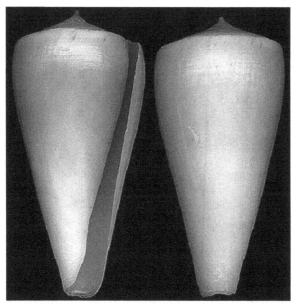

Apertural view   Abapertural view

Common Name: Not designated

Geographical Distribution: SE Africa – Oman; Mascarenes

Habitat: Between 60–143 m; sandy substratum

Identifying Features: Shell of this species is obconic in shape with broad shoulders. Protoconch is paucispiral. Whorl tops have cords, and there is a well-developed dentiform plait. Anal notch is shallow to relatively deep, and an anterior notch is absent. Periastracum is smooth and the operculum is large. Anterior section of the radular tooth is shorter than the posterior section. Barb is short. There are one or two rows of serrations. It grows to a maximum size of 41 cone snails are presumed vermivorous, preying on polychaete worms. These snails are predatory and venomous. They are capable of "stinging" humans and therefore live ones should not be handled.

Toxin Type: Conopeptides; venom type not reported.

Conus andremenezi (Olivera & Biggs, 2010) (= *Kuradoconus andremenezi*)

Abapertural view   Apertural view

Common Name: Andremenez's Cone

Geographical Distribution: West Pacific; throughout the central Philippines, from Cebu City south to Aliguay Islands

Habitat: Depths of between 180–300 m on mud and gravel

Identifying Features: Shell of this species has a biconical in shape and moderately solid with a relatively high spire. Last whorl is broadly conical, with raised spiral ribs, which are not smooth but always undulating (and in some specimens, the ribs seem to have arch-like protuberances, instead of a continuous smooth rib). Raised ribs on the body whorl are well separated from each other, with interstices that have axial scales between them. Body whorl has an off-white ground color with characteristic purplish-brown spots that occur in zones. In two darker zones, spots generally cover more spiral ribs and extend into the interspaces. Spots begin to appear on the periphery of the third or fourth teleoconch whorl, and typically these are more closely spaced to each other than are the larger spots in the later spire whorls. Adults may reach a maximum size of 53 mm. Like all species within the genus Conus, these snails are predatory and venomous. They are capable of "stinging" humans and therefore live ones should not be handled.

Toxin Type: Conopeptides; venom type not reported.

Conus antoniomonteiroi **(Rolán, 1990)**

Apertural view   Abapertural view

Common Name: Antonio Monteiro's cone.

Geographical Distribution: Atlantic Ocean off the Cape Verde Archipelago.

Habitat: Under rocks and stones in very shallow water; more abundant in some sheltered small bays

Identifying Features: This species displays a characteristic profile consisting of an elevated, concave spire and a well-marked shoulder. Color of the shell is greenish brown to olive green, having a white band in the midbody of the last whorl overlaid with dark brown marks. A similar, narrower band is also present just below the shoulder. This species grows to a maximum size of 20 mm only. Like all species within the genus Conus, these snails are predatory and venomous. They are capable of "stinging" humans and therefore live ones should not be handled.

Toxin Type: Conopeptides; venom type not reported.

Conus apluster **(Reeve, 1843)** (*=Floraconus apluster*)

Apertural view Axial view Abapertural view

Common Name: Little Flag cone, back-end cone

Geographical Distribution: New South Wales and S. Queensland, Australia.

Habitat: Under rocks on exposed rocky shores, at low tide level.

Identifying Features: Shell of this species is small to moderately small or light to moderately solid. Last whorl is ventricosely conical to broadly and ventricosely conical. Outline is convex adapically and straight toward base. Left side may be concave near base. Shoulder is angulate to rounded. Spire is of low to moderate height and outline is straight to sigmoid. Teleoconch sutural ramps are flat to slightly concave. Last whorl is with rather widely spaced distinct spiral ribs on basal third to half. Ground color is grayish to light blue. Last whorl is with 2–5 (usually 3) variably broad, olive to brown or pink spiral bands. Variably spaced spiral rows of small to large squarish brown dots are seen extending from base to shoulder. This species has a maximum length of 33 mm. Like all species within the genus Conus, these snails are predatory and venomous. They are capable of "stinging" humans and therefore live ones should not be handled.

Toxin Type: Conopeptides; venom type not reported.

Conus ardisiaceus (Kiener, 1845)

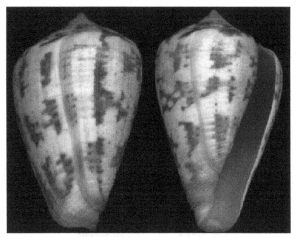

Abapertural view   Apertural view

Common Name: Not designated

Geographical Distribution: Muscat to Al Masirah, Oman

Habitat: Shallow water; on coral reef.

Identifying Features: Shells of this species are moderately small to medium-sized, and usually moderately solid. Last whorl is ventricosely conical to broadly and ventricosely conical, sometimes broadly ovate. Outline is convex and less so to straight toward base. Left side is constricted at base or sigmoid. Shoulder is rounded to angulate. Spire is of low to moderate height and its outline is concave to almost straight or slightly sigmoid. Last whorl is with a few usually weak spiral ribs at base. Ground color is white to light grayish blue. Last whorl is usually with brown, olive, orange, or blackish brown flecks, variable in shape and arrangement and often fusing into 3 spiral bands, at center, below shoulder and at base, and with axial streaks or blotches. Spiral rows are of variably alternating brown and white to light gray dots and dashes generally extending from base to shoulder. Apex is white to orange. Aperture is light to dark violet. Size of the shells varies between 25 and 55 m. Like all species within the genus Conus, these snails are predatory and venomous. They are capable of "stinging" humans and therefore live ones should not be handled.

Toxin Type: Conopeptides; venom type not reported.

Rolaniconus athenae **(Filmer, R.M., 2011) (=** *Conus athenae***)**

Apertural view

Common Name: Not designated

Geographical Distribution: Hawaii

Habitat: Coral and mud bottom

Identifying Features: Shell of this small and is ventricosely conical to conoid-cylindrical in shape, with a moderate to high spire of straight outline. There are six to nine turreted spire whorls below the protoconch. Sutural ramps have a slightly concave profile. Early four spire whorls are beaded and contain a strong spiral cord on the inner edge at the suture and a weaker one on the outer edge. The following five spire whorls are strongly nodulose at the outer edge and are concave in outline. Body whorl is very slightly convex in outline. It is covered from base to shoulder with strong and regular spiral cords which are rounded and evenly nodulose. Body whorl has a dull shine and is white or very faint yellow-tan with no distinguishing marks. The size of the shell varies from only 13.6 to 22.7 mm in length. Like all species within the genus Conus, these snails are predatory and venomous. They are capable of "stinging" humans and therefore live ones should not be handled.

Toxin Type: Conopeptides; venom type not reported.

Conus attenuatus **(Reeve, 1844)** (= *Attenuiconus attenuatus*)

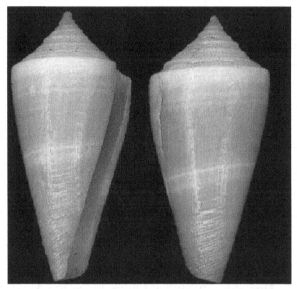

Apertural view   Abapertural view

Common Name: Slender cone

Geographical Distribution: From Colombia to Bahamas; south-east Florida and Florida Keys; through Central America; in all West Indies islands; and along northern South American coasts.

Habitat: Sand and rubble substrate on coral reefs at depths of between 5 and 20 m.

Identifying Features: Shell of this species is very elongated, with straight sides and narrow, straight apertures. Spires are low or flattened, with projecting, mammilate protoconchs of 2 or 3 whorls. Spire whorls may be flattened, or slightly canaliculated. Shells are generally smooth and polished. Shells are generally colored in yellows or oranges arranged in wide bands, but may be colored pink, salmon, reddish-orange with brown or white longitudinal flammules. Adults of the species will grow to 28 mm. Like all species within the genus Conus, these snails are predatory and venomous. They are capable of "stinging" humans and therefore live ones should not be handled.

Toxin Type: Conopeptides; venom type not reported.

Vituliconus augur **(Lightfoot, J., 1786) (=** *Conus augur***)**

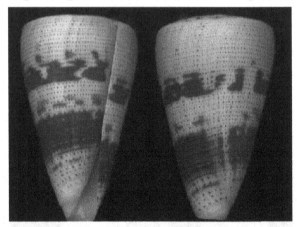

Apertural view   **Abapertural view**

Common Name: Auger cone

Geographical Distribution: Indian Ocean along the Aldabra Atoll and Madagascar; and in the South-west Pacific Ocean

Habitat: In 3–25 m; living in muddy sand, on coral rubble and beneath rocks

Identifying Features: Shell of this species is medium sized to large and solid to heavy. Body whorl is broadly conical and sides are nearly straight. Shoulder is subangulated weakly tuberculate. Spire is of moderate height and its outline is moderately convex. Body whorl is with weak spiral ribs at base and ribs are granulose in large specimens. Ground color is white to pale yellow. Body whorl is with numerous spiral rows of fine reddish brown dots from base to shoulder, with two interrupted reddish brown transverse bands on either side of the center. Posterior band extends irregularly towards the shoulder. Aperture is white, and outer lip is thick. Size of the shell varies from 50 to 80 mm. Like all species within the genus Conus, these snails are predatory and venomous. They are capable of "stinging" humans and therefore live ones should not be handled.

Toxin Type: Conopeptides; venom type not reported.

*Conus aurantius (*Hwass in Bruguière, 1792) (= *Tenorioconus aurantius*)

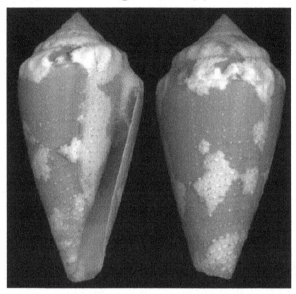

Apertural view   Abapertural view

Common Name: Golden cone

Geographical Distribution: Tuamotu Archipelago in French Polynesia

Habitat: Inter-reef substrate at depths up to 10 m

Identifying Features: Shell is elongtely cylindrical and smooth. Spire is conical and obtuse at apex. There are eight whorls, which are obliquely flattened on top with channeled sutures. Entire surface is decorated with reddish-brown blotches of varying size surrounded by heart-shaped or triangular rosy-pink spots. Aperture is wide a broadened at anterior end. Adults will grow to 70 mm. Like all species within the genus Conus, these snails are predatory and venomous. They are capable of "stinging" humans and therefore live ones should not be handled.

Toxin Type: Conopeptides; venom type not reported.

Asprella baeri (Röckel and Korn, 1992) (= *Conus baeri*)

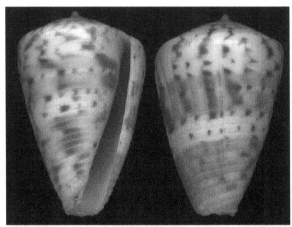

Apertural view   **Abapertural view**

Common Name: Baer's cone

Geographical Distribution: Southern Mozambique

Habitat: Most probably it occurs on mud from depths of about 450 m

Identifying Features: Shell of this species is moderately small to medium-sized and moderately solid. Last whorl is usually conical to ventricosely conical and its outline is convex adapically and straight toward base. Shoulder is angulate. Spire is of low to moderate height and it s outline is almost straight to slightly concave. Teleoconch sutural ramps are nearly flat, with 2 increasing to 3–5 spiral grooves. Last whorl is with distinct spiral grooves from base to center or shoulder and ribbons between. Ground color is white to pale orange. Last whorl is usually with a light orangish brown spiral band above and below center, occasionally with an additional smaller band below shoulder. 10–15 spiral rows of reddish brown spots or bars are found extending from base to shoulder, sometimes fusing into irregular axial markings. Aperture is brownish cream to pale orange. Shell size varies from 30 to 45 mm. Like all species within the genus Conus, these snails are predatory and venomous. They are capable of "stinging" humans and therefore live ones should not be handled.

Toxin Type: Conopeptides; venom type not reported.

Conus balteatus **(G. B. Sowerby I, 1833) (=***Rolaniconus balteatus***)**

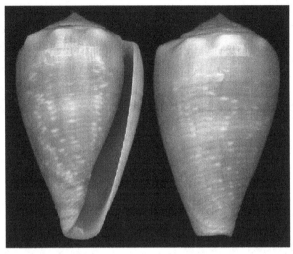

Apertural view   Abapertural view

Common Name: Ringed cone

Geographical Distribution: Indian Ocean: Mozambique to N. Somalia, Mascarenes, Maldives and Indonesia; Pacific: Japan to W. Australia and Queensland and to Fiji and Samoa.

Habitat: Intertidally and slightly subtidally on coral reef platforms, living on rough limestone, dead coral rocks, rubble and rubble mixed with sand, often hidden beneath coral rocks.

Identifying Features: Shell of this species is small to medium-sized and is moderately light to moderately solid. Last whorl is conical or ventricosely conical to broadly conical or pyriform and outline is slightly to distinctly convex adapically, less so, straight or somewhat concave below. Shoulder is usually angulate, and is strongly to weakly tuberculate. Spire is of low to moderate height, and its outline is straight to concave. Teleoconch sutural ramps are flat, with 1–2 increasing to 4–7 spiral grooves. Entire last whorl is with closely spaced spiral ribs, weak in some populations. Ground color is white, sometimes bluish violet. Last whorl is encircled with a color band, of various shades of brown to brownish red or olive, on each side of center. Dark zones of last whorl is often speckled with white dots or dashes arranged in spiral rows, either scattered or regularly arrayed. Aperture is violet to brown in larger shells. Shell size varies from 25 to 47 mm. Like all species within the genus Conus, these snails are predatory and venomous. They are capable of "stinging" humans and therefore live ones should not be handled.

Toxin Type: Conopeptides; venom type not reported.

Conus (Purpuriconus) baiano (Coltro, 2004) (= *Conus ziczac*)

Apertural view   **Abapertural view**

Common Name: Bahia State cone

Geographical Distribution: Southern Bahia State, Brazil

Habitat: Rubble and coral sand bottom at 10–25 m on offshore reefs.

Identifying Features: This species has a small to medium-sized shell. Spire is concave-sided and elevated. Shoulder of the body whorl is smooth. Body whorl is slightly convex with 6–8 incised lines on the base. Apex is pink-white to white. Spire is with 5–7 whorls, with medium deep suture with white and brown dots, each whorl with 3–5 distinct spiral ridges crossed by many fine curved axial threads. Color of the body whorl is bright red with white marks and brown dots lines, sometimes dark purple-brown and white. Pink red aperture is seen on red specimens or purple aperture on purple specimens. Shell has the maximum size of 33 mm. Like all species within the genus Conus, these snails are predatory and venomous. They are capable of "stinging" humans and therefore live ones should not be handled.

Toxin Type: Conopeptides; venom type not reported.

Conus barbieri **(Raybaudi Massilia, 1995)**

Apertural view   Axial view   Abapertural view

Common Name: Not designated

Geographical Distribution: Throughout the Philippines

Habitat: Found inside giant barnacles at depths of 25 m.

Identifying Features: This species has an elongated, cylindrical and glossy shell. Fine incised spiral threads from shoulder to base; early 4–5 postembryonic whorls tuberculated. Protoconch and early 5 teleconch whorls are white. Aperture is bluish-white and periostracum is smooth, dull, brown. Ground color is white or grayish-blue, but with brown general appearance from network of axial and spiral lines, leaving small, white, tent markings of variable size, often forming bands. Adults can grow up to 32 mm. Like all species within the genus Conus, these snails are predatory and venomous. They are capable of "stinging" humans and therefore live ones should not be handled.

Toxin Type: Conopeptides; venom type not reported.

Conus barthelemyi (Bernardi, M., 1861)

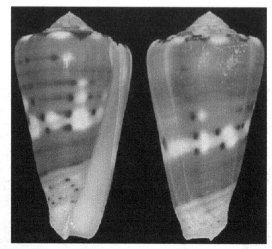

Apertural view   Abapertural view

Common Name: Barthelemy's cone

Geographical Distribution: Off islands throughout the Indian Ocean, ranging from Réunion to the Maldives and Seychelles; Christmas Island

Habitat: In 10–30 m, inhabiting sand and rock substrata

Identifying Features: Shell of this species is medium-sized to large and moderately solid-to-solid. Last whorl is ventricosely conical-to-conical and outline is convex at adapical fourth, straight below, occasionally slightly concave centrally. Shoulder is broadly carinate to carinate. Spire is low and is of moderate height and its outline is usually sigmoid, sometimes nearly straight. Teleoconch sutural ramps are almost flat, grading to deeply concave in late whorls, with 2 increasing to 4–9 spiral grooves that are often weak on latest ramps. Ground color is white. Last whorl is with variably broad, orangish to reddish or violet brown spiral bands usually leaving interrupted to solid narrow white bands centrally and at base. Spiral rows of blackish brown dots, spots, bars, and blotches are seen from base to shoulder. Aperture is white to bluish white. Shell size varies from 42 to 84 mm. Like all species within the genus Conus, these snails are predatory and venomous. They are capable of "stinging" humans and therefore live ones should not be handled.

Toxin Type: Conopeptides; venom type not reported.

Conus bayani **(Jousseaume, 1872)** **(=** *Stellaconus bayani*)

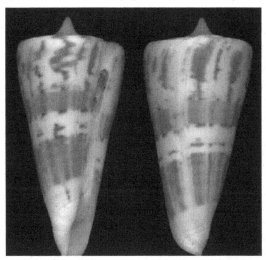

Apertural view   Abapertural view

Common Name: Bayan's cone

Geographical Distribution: Indian Ocean; (north-west Indian Ocean from the central Red Sea to Somalia)

Habitat: In depths of 20–100 m, mainly on sand

Identifying Features: Shell of this species is medium sized to moderately large, and moderately solid. Body whorl is conical and outline is straight to slightly sigmoid. Shoulder is broad and carinate. Body whorl is with distinct or weak spiral ribs and ribbons at base separated by fine grooves. Spire is of moderate height, extremely concave and sharply pointed. Aperture is generally narrow and outer lip is straight, sharp, thin and fragile. Ground color is white to pinkish or cream. Body whorl is variously covered with deep yellowish-brown blotches and streaks. Spire whorls are with scattered blotches of brown. Aperture is white, with violet tone. Size of an adult shell varies between 45 mm and 70 mm. The species has planktotrophic larval development, which explains its widespread distribution. Like all species within the genus Conus, these snails are predatory and venomous. They are capable of "stinging" humans and therefore live ones should not be handled.

Toxin Type: Conopeptides; venom type not reported.

Cylinder bengalensis **(Okutani, 1968)** (= *Conus (Cylinder) bengalensis*)

Apertural view   Abapertural view

Common Name: Bengal cone

Geographical Distribution: Andaman Sea, Bay of Bengal, S.E. India; Red Sea

Habitat: In 50–130 m, on mud and sand bottoms

Identifying Features: Shell of this species is large, solid and tall with high gloss. Body whorl is narrowly conical and outline straight. Shoulder is narrow, straight angled, slightly concave above and merging into spire. Spire is of moderate height and stepped. Its outline is almost straight and sharply pointed. Aperture is narrow, parallel, and is slightly wider posteriorly. Anal notch is very deep and outer lip is sharp, straight and is slightly curved anteriorly. Ground color is white and is heavily covered with dense network of small and medium reddish-brown tents. Two rows of dark reddish brown blotches, one above the center and the other below are seen. Spire is with similar reddish-dark brown blotches. Spire tip is pale pinkish. Aperture is white Maximum reported size of this species is 154 mm. It feeds on gastropods. Like all species within the genus Conus, these snails are predatory and venomous. They are capable of "stinging" humans and therefore live ones should not be handled.

Toxin Type: Conopeptides; venom type not reported.

Conus biliosus **(Röding, 1798) (=***Lividoconus biliosus***)**

Apertural view   Abapertural view

Common Name: Meyer's Bilious cone

Geographical Distribution: Western Indian Ocean (from South Africa to Somalia) and along India and Sri Lanka; Pacific Ocean from Indonesia to Philippines and to Papua New Guinea; Solomon Islands and Queensland

Habitat: Intertidal and subtidal habitats

Identifying Features: Shell of this species is moderately small to large and solid with a low gloss or dull finish. Outline is conical and sides are straight or inflated posteriorly. Body whorl usually elongates and is covered with numerous low and undulating spiral ridges from base to shoulder. Shoulder is wide, roundly angled, weakly coronated or undulate, and is usually slightly concave above. Spire is low and bluntly pointed and its sides are straight to slightly concave. Aperture is moderately wide and uniform. Outer lip is nearly straight, moderately thick and sharp. Ground color of shells is pearly white with pale dashes. Shell size varies from 24 to 55 mm. These snails are vermivorous preying on a broad variety of polychaete worms, including enteropneust, eunicid, terebellid, cirratulid, maldanid, and nereid worms, and hemichordates. Like all species within the genus Conus, these snails are predatory and venomous. They are capable of "stinging" humans and therefore live ones should not be handled.

Toxin Type: Conopeptides; venom type not reported.

Conasprella biraghii (G. Raybaudi, 1992) (= *Conus biraghii*)

Apertural view   **Abapertural view**

Common Name: Biraghi's cone

Geographical Distribution: Southern Red Sea, western Arabian Sea and the Gulf of Oman; coast of Somalia from Mogadishu to Obya

Habitat: There are no reliable data on the habitats and ecology of this species

Identifying Features: Shell of this species is biconic, slender and rather solid. Protoconch is of 1.5 whorls, initially mostly white with brown sutures, then brown with minute axial folds. Body whorl is smooth, with a groove just below the shoulder. Shell coloration is white, with a grayish upper band ornamented with spiral white/brown lines. A second grayish band, equally ornamented, also occurs near the base. Between the two bands there are spiral lines spotted with brown and white. Adults of the species will grow to

12 mm. Like all species within the genus Conus, these snails are predatory and venomous. They are capable of "stinging" humans and therefore live ones should not be handled.

Toxin Type: Conopeptides; venom type not reported.

Conus blanfordianus (Crosse, 1867)

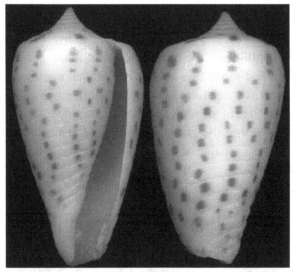

Apertural view   **Abapertural view**

Common Name: Black and white cone

Geographical Distribution: Pacific Ocean along Philippines and Papua New Guinea

Habitat: Shallow water at depths of 20–100 m; muddy areas

Identifying Features: Light-colored or dark colored species with moderately small, and moderately light to moderately solid shell. Last whorl is conoid-cylindrical to ventricosely conical. Outline convex is at adapical third and less so below. Aperture is wider below center than near shoulder. Shoulder is angulate to subangulate. Spire is of low to moderate height and outline is deeply concave. Teleoconch sutural ramps are flat, with 1–2 increasing to 4 wide spiral grooves. Last whorl is with axially striate spiral grooves from base to center; intervening ribbons grade to ribs near base.

Dark-colored form is medium sized and moderately solid. Last whorl is more conical than in other form. Shoulder is angulate. Spire is of moderate height and outline is concave, with late spire whorls more raised than in light-colored form. Teleoconch sutural ramps are with 1 increasing to 6–8 spiral grooves and ribs between are variably broad. Last whorl is with spiral grooves to center or shoulder. Ground color is white. In light-colored shells, last whorl is with about 15 spiral rows of brown spots that may fuse axially. Aperture is white. In dark-colored shells, spiral rows consist of larger dark brown spots and bars. Markings are sparse within a spiral band below center. Additional fine brown dots and axial dashes are seen on subcentral band and sometimes at shoulder. Radial markings of ramps are large and dark brown. Aperture is white, becoming orange deep within in large specimens. Size of an adult shell varies between 22 mm and 58 mm. Like all species within the genus Conus, these snails are predatory and venomous. They are capable of "stinging" humans and therefore live ones should not be handled.

Toxin Type: Conopeptides; venom type not reported.

Conus (Africonus) boavistensis **(Fernandes, 1990) (=** *Africonus boavistensis*)

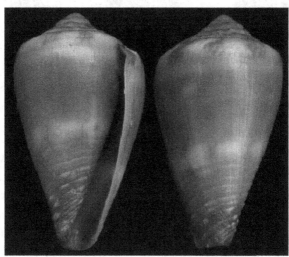

Apertural view   Abapertural view

Common Name: Boavista cone

Geographical Distribution: Atlantic Ocean along the Cape Verde islands.

Habitat: At rocks in shallow waters

Identifying Features: This species has a very small, wide shell with a slightly angulated shoulder. Spire is moderately high and sutural ramps straight with spiral striae. Shell color is bluish-gray to greenish, with two bands of light bluish dashes (one below the central portion of the last whorl and the other between the first one and the shoulder). Dashes are of irregular shape, but they usually form oblique or angular lines. Size of an adult shell grows to a length of 13–20 mm. Aperture is dark bluish, with two light bands. Like all species within the genus Conus, these snails are predatory and venomous. They are capable of "stinging" humans and therefore live ones should not be handled.

Toxin Type: Conopeptides; venom type not reported.

Conus (Jaspidiconus) bodarti **(Coltro, 2004)** (= *Conasprella ians*)

Apertural view   Abapertural view

Common Name: Not designated

Geographical Distribution: Bahia State, Brazil

Habitat: On rubble and coral sand bottom at 20–25 m, limited to off-shore reefs

Identifying Features: Shell of this species has convex sides in adult specimens. There is a straight-sided spire. Shoulder is roundly angulated and nodulose. Body whorl is with 12–14 incised lines, starting near the siphonal canal up to middle of the body. Apex is yellowish smooth with 2 to 2-1/5 whorls. Spire is with 6 to 8 whorls, with medium deep suture. Color of body is red-brown with gray and white marks. 18–20 spiral cords with interrupted brown and white dots are seen. Aperture is white. Maximum reported size is 19.2 mm. Like all species within the genus Conus, these snails are predatory and venomous. They are capable of "stinging" humans and therefore live ones should not be handled.

Toxin Type: Conopeptides; venom type not reported.

Rolaniconus boeticus (Reeve, L.A., 1844)

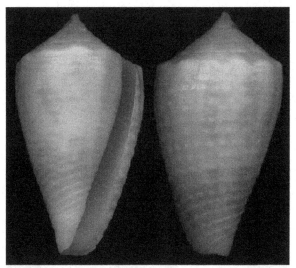

Apertural view   Abapertural view

Common Name: Boeticus cone

Geographical Distribution: Indian Ocean along Mozambique, the Seychelles and the Mascarene Basin and in the Pacific Ocean along Japan, Indonesia, Fiji and Australia

Habitat: Seamounts and knolls

Identifying Features: Striate spire of this species is slightly tuberculate. Body whorl is granular, and striate towards the base. Color of the shell is white, marbled with chestnut or chocolate and with revolving rows of chestnut spots. Shell size varies from 15 to 40 mm. Like all species within the genus Conus, these snails are predatory and venomous. They are capable of "stinging" humans and therefore live ones should not be handled.

Toxin Type: Conopeptides; venom type not reported.

Stellaconus bondarevi **(Röckel, D., and Raybaudi Massilia, G., 1992) (=** *Conus bondarevi***)**

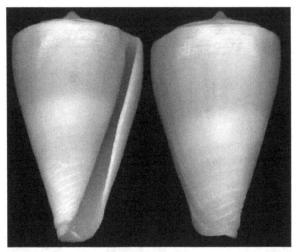

Apertural view   Abapertural view

Common Name: Bondarev's cone

Geographical Distribution: Endemic to the coast of northern Somalia at the Horn of Africa

Habitat: Deep water species at depths of about 120–150 m

Identifying Features: Shell of this species is moderately small to medium-sized and moderately solid. Last whorl is conical to broadly conical. Outline is convex below shoulder and straight below. Shoulder is angulate. Spire is low and outline is sigmoid to concave. Teleoconch sutural ramps are flat, with 2 increasing to 3–4 spiral grooves. Last whorl is with distinct spiral ribs on basal fourth to third. Ground color is white, sometimes suffused with vale orange or violet. Last whorl is with a broad yellow, orange, red or brown spiral band on each side of center, occasionally with a narrow third band below shoulder. Base is white to orange. Aperture is white to pale purple and slightly translucent. Shell size varies from 29 to 43 mm. Like all species within the genus Conus, these snails are predatory and venomous. They are capable of "stinging" humans and therefore live ones should not be handled.

Toxin Type: Conopeptides; venom type not reported.

Conus bonfigliolii (Bozzetti, 2010) (= *Conus bozzetti*)

Apertural view   Abapertural view

Common Name: Not designated

Geographical Distribution: This is a newly described species and is currently known to occur around Lavanono, in South Madagascar

Habitat: There is currently no habitat information for this species; probably lives at depths of 30–50 m

Identifying Features: Shell of this species is chalky white, light and thin. Spire at times is dotted with chestnut brown. Body whorl is smooth and moderately glossy, only the base is ridged with 5 to 8 oblique ribs.

Protoconch is rather broad and low, pure white, with 1.5–2 whorls. Spire whorls form a heightened slope on their external periphery, which is distinctly nodulose on the 5 or 6 earliest whorls. It can grow to 27–28 mm in size. Like all species within the genus Conus, these snails are predatory and venomous. They are capable of "stinging" humans and therefore live ones should not be handled.

Toxin Type: Conopeptides; venom type not reported.

***Conus boschorum* (Moolenbeek and Coomans, 1993) (= *Pseudolilliconus boschorum*)**

Apertural view   **Abapertural view**

Common Name: Not designated

Geographical Distribution: Endemic to Oman where it is restricted to the waters off Masirah Island

Habitat: Shallow water species at depths of 0.1–6 m.

Identifying Features: Shell of this species is thin and glossy. Protoconch is with 1.5 whorls. Spire is stepped and whorls are canaliculated. Shoulder is sharply angulated. Body whorl is smooth except for lower third, which has spiral grooves. Spire is white, with irregular dark brown spots. On the body whorl there are 8 brown spots, which continue below the shoulder and are connected to a broader blackish band. Like all species within the genus Conus, these snails are predatory and venomous. This is a small cone snail with adults of the species growing to approx. 13 mm. They are capable of "stinging" humans and therefore live ones should not be handled.

Toxin Type: Conopeptides; venom type not reported.

Conus brianhayesi **(Korn, 2002) (=** *Conus hayesi***)**

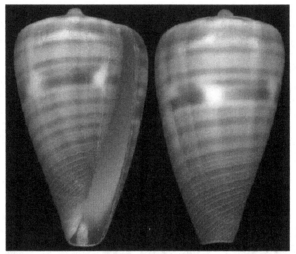

Apertural view   Abapertural view

Common Name: Brian Haye's cone

Geographical Distribution: Found in a very localized area, in the northern Transkei region of South Africa.

Habitat: Deep-water species at depths of about 70 to 100 m

Identifying Features: Protoconch of the species is about 1.75 whorls and its maximum diameter is 1.5–1.6 mm. Ground color of the shell is white to cream and base is stained with brown or violet brown. Basic color pattern of last whorl is composed of brown spiral elements. Spiral lines and/ or spiral rows of dashes are found extending from base to shoulder. Shell is up to 25 mm in length. Like all species within the genus Conus, these snails are predatory and venomous.

Toxin Type: Conopeptides; venom type not reported.

Conus broderipii (Reeve, 1844) (= *Graphiconus broderipii*)

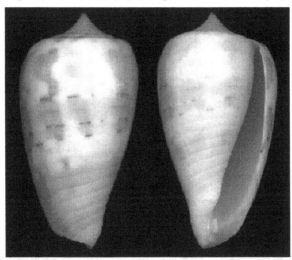

Abapertural view   Apertural view

Common Name: Broderip's cone

Geographical Distribution: Philippines to the Moluccas (Indonesia)

Habitat: Shallow-subtidal zone to about 20 m.

Identifying Features: Shells of this species are moderately small and moderately light. Last whorl is ventricosely conical to sometimes conoid-cylindrical. Outline is convex adapically, less so to straight below. Aperture is wider at base than near shoulder. Shoulder is angulate to sometimes subangulate. Spire is low and outline is concave. Teleoconch sutural ramps are flat, with 1 increasing to 2–3 major spiral grooves. Last whorl is with broad spiral ribbons below center, narrow or replaced by ribs within basal third. Ground color is white. Last whorl is with spirally aligned light to reddish brown dots, spots and bars concentrated or fused into 3 spiral bands, below shoulder, above center and within basal third. Aperture is violet to light purple, with or without a darker collabral band. Adults typically grow to 30 mm in length. Like all species within the genus Conus, these snails are predatory and venomous.

Toxin Type: Conopeptides; venom type not reported.

Conus   cakobaui   (Moolenbeek,   Röckel   and   Bouchet,   2008)
(= *Profundiconus cakobaui*)

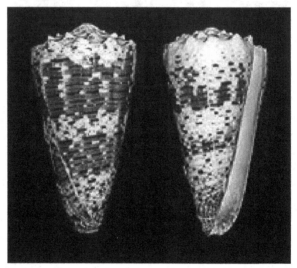

Abapertural view and vice-versa   Abapertural view

Common Name: Not designated

As it is a newly discovered species, nothing much is known about its biology.

Identifying Features: Shell of this species is small, thin and narrowly conical. Protoconch is smooth, paucispiral and of 1.5 convex whorls with a diameter of 1025 μm. Teleoconch is of 7.5 whorls with rather deep suture. Spire is rather high, and its profile is nearly flat. Shoulder is angulate. First 3 teleoconch whorls with fine tubercles, which are gradually disappearing on subsequent whorls. Last whorl is with 4 fine spiral grooves on periphery and about 15 on the base. Color of protoconch is transparent white. First teleoconch whorls are creamy white with a brown spiral band on the periphery, extending over the row of tubercles or just adapically of it. On later whorls, this brown band is interrupted by white areas. Last whorl is white with an irregular brown pattern. Tip of base is white. Height and width of the shell are 18.9 mm and 8.4 mm, respectively. Like all species within the genus Conus, these snails are predatory and venomous.

Toxin Type: Conopeptides; venom type not reported.

Kioconus capreolus (Röckel, D., 1985) (= *Conus capreolus*)

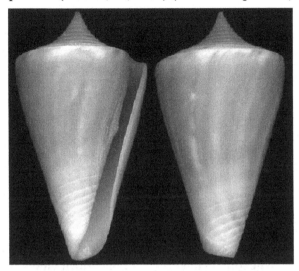

Apertural view   **Abapertural view**

Common Name: Not designated

Geographical Distribution: E. India to Andaman Sea.

Habitat: Reported from about 50 m.

Identifying Features: Shell of this species is medium-sized to moderately large and moderately solid. Last whorl is conical and outline is almost straight. Large shells are with a concave right side. Shoulder is sharply angulate. Spire is of moderate height and early whorls are stepped. Teleoconch sutural ramps are flat to slightly concave, with pronounced closely set radial threads Last whorl is with a few spiral grooves at base, separated by ribs anteriorly and by ribbons posteriorly. Ground color is white, suffused with violet and cream on last whorl. Last whorl is with fawn axial streaks, varying from separate to fused in an almost solid brown coloration. Larval whorls are white. Aperture is cream white, shaded with very pale violet. Shell size varies from 36 to 65 mm. Like all species within the genus Conus, these snails are predatory and venomous.

Toxin Type: Conopeptides; venom type not reported.

Conus cardinalis **(Hwass in Bruguière, 1792)** (= *Conus (Purpuriconus) donnae*)

Apertural view   **Abapertural view**

Common Name: Cardinal cone

Geographical Distribution: North-westernmost Great Bahama Bank, Bahamas; Honduras, Roatan Island

Habitat: Shallow waters of 0–21 m depths

Identifying Features: Shell of this species is with high, stepped spire and sharply angled shoulder, which is weakly coronated and ornamented with low undulating knobs. Shell color is normally bright yellow-orange. Maximum-recorded shell length is 32.2 mm. Like all species within the genus Conus, these snails are predatory and venomous.

Toxin Type: Conopeptides; venom type not reported.

Conus ceruttii (Cargile, 1997)

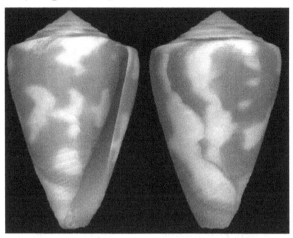

Apertural view   **Abapertural view**

Common Name: Not designated

Geographical Distribution: Caribbean coastline of Honduras and islands of Nicaragua (Corn Islands) and Colombia

Habitat: Depths of 10–25 m probably in coral sand

Identifying Features: Protoconch of this species is with two smooth nuclear whorls above two weakly tuberculated early teleoconch whorls. Juvenile shells (<20 mm) are pink, violet, orange or yellow, with an irregular white band in middle of body whorl. Adult shells are bright reddish-orange with a more defined white band. Maximum-recorded shell length is 40.3 mm. Like all species within the genus Conus, these snails are predatory and venomous.

Toxin Type: Conopeptides; venom type not reported.

Conus chaldeus (Röding, 1798)

Abapertural view   Apertural view

Common Name: Vermiculated cone, Astrologer's cone, Small spots Conus

Geographical Distribution: Indo-Pacific (Kenya)

Habitat: Intertidal waters on rocky reef platforms and coral reef rims; sandy and muddy substrata

Identifying Features: Shell of this species is small with dark brown wavy lines with two white bands. Shells vary in size from 15.5 and 26 mm. Like all species within the genus Conus, these snails are predatory and venomous.

Toxin Type: Conopeptides; venom type not reported.

Asprella ciderryi (**Motta, A.J. da, 1985**) (= *Conus ciderryi*)

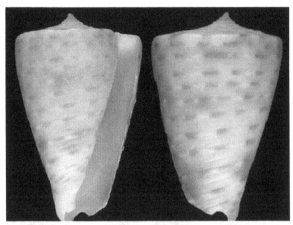

Apertural view   Abapertural view

Common Name: Not designated

Geographical Distribution: Taiwan and Vietnam; Philippines; Amami Islands

Habitat: Deeper subtidal zones at about 200 m; sandy or muddy substratum

Identifying Features: Shell of this species is moderately small to medium-sized and moderately light to moderately solid. Last whorl is onical and outline is almost straight. Shoulder is angulate, undulate or weakly tuberculate. Spire is of low to moderate height and outline is concave. Teleoconch sutural ramps are flat, with 1–2 increasing to 2–5 spiral grooves. Last whorl is with rather widely spaced spiral grooves on basal half, more narrowly spaced near base. Ground color is white, occasionally suffused with pale violet. Entire last whorl is with spiral rows of alternating yellowish brown or pink and white bars, squarish spots and dots. An orangish violet spiral band or a spiral row of yellowish brown-to-brown flecks may also occur on each side of center. Aperture is white. Shell size varies from 30 to 42 mm. Like all species within the genus Conus, these snails are predatory and venomous.

Toxin Type: Conopeptides; venom type not reported.

Conus cinereus **(Hwass in Bruguière, 1792)**

Abapertural view **Apertural view**

Common Name: Sunburnt cone

Geographical Distribution: Western Pacific Ocean from Japan to Indonesia

Habitat: Upper subtidal zone

Identifying Features: Shell of this species is cylindrically ovate, with a moderate, smooth spire. Body whorl is encircled below by distant grooves. Shell is clouded with olivaceous, ashy blue and chestnut-brown, with revolving lines articulated of chestnut and white spots. Brown-stained aperture is wider at its base than at its shoulder. Size of an adult shell varies between 15 mm and 57 mm. Like all species within the genus Conus, these snails are predatory and venomous.

Toxin Type: Conopeptides; venom type not reported.

Conus circumactus **(Iredale, 1929) (=** *Vituliconus circumactus***)**

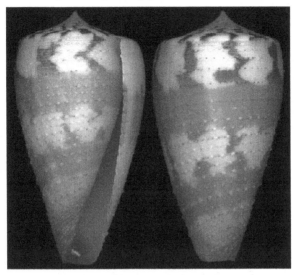

Apertural view   Abapertural view

Common Name: Circumactus cone

Geographical Distribution: Pacific Ocean and along Australia

Habitat: At depths between 10–240 m

Identifying Features: This species has a smooth shell, which is rather thin. Spire is low-conical and contains revolving striae, usually maculated with chestnut. Body whorl is striate below. Color of the shell is yellowish or light chestnut, with large white blotches forming a band at the shoulder and another on the middle, encircled by narrow chestnut lines, which are often broken up into small dots. Color of the base and the aperture is usually violaceous. Size of an adult shell varies between 35 mm and 75 mm. Like all species within the genus Conus, these snails are predatory and venomous.

Toxin Type: Conopeptides; venom type not reported.

Conus claudiae (Tenorio and Afonso, 2004)

Abapertural view and vice-versa   Abapertural view

Common Name: Not designated

Geographical Distribution: Praia Real and Baía de Navío Quebrado, North of Maio Island, Cape Verde Island.

Habitat: At 1–2 m depth; attached to the bottom of rocks and dead coral slabs on sandy bottom; occasionally in crevices of rocky platforms, with algae and sand cover.

Identifying Features: This species has a small to moderately small shell. Profile is conical to broadly conical, with a moderate spire and a subangulated shoulder. Outline of the last whorl is convex. Spire is most often eroded, concave with flat or slightly convex sutural ramps presenting fine striae. Ground color of the shell varies from pale yellow-green to light bluish gray. There is a reticulated pattern of white flecks and dark brown dots forming bands, which are variable in number and width. Usually there are three such bands viz. a thin one on the shoulder, another at the height of the maximum diameter of the shell, and another broader one slightly below the midbody. Base is dark, often covered by reticulated pattern of white flecks and dark brown dots forming bands, which are variable in number and width. Base is dark, often covered by reticulated pattern. Aperture is purplish brown, with two white bands, one in the middle portion and another one in the upper part. Inner lip is white, showing traces of the yellow near the edge. Shell size varies from 16 to 26 mm. Like all species within the genus Conus, these snails are venomous to humans.

Toxin Type: Conopeptides; venom type not reported.

Conus coccineus (Gmelin, 1791) (= *Rolaniconus coccineus*)

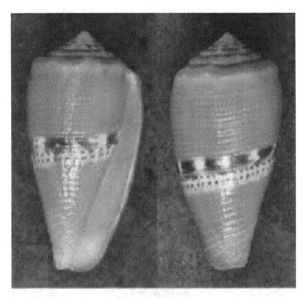

Apertural view   Abapertural view

Common Name: Scarlet cone, berry cone

Geographical Distribution: E. Indonesia, Philippines, Queensland, New Caledonia, Solomon Island, and Vanuatu.

Habitat: In 1–20 m, at exposed coral reef sites and in coral rubble

Identifying Features: Shell of this species is moderately small to moderately large or moderately solid. Last whorl is ovate to conoid-cylindrical and sometimes cylindrical. Outline is convex to almost straight and parallel-sided adapically. Left side is concave basally. Shoulder is angulate and undulate to weakly tuberculate. Spire is of low to moderate height. Early postnuclear whorls are tuberculate and late whorls are tuberculate to undulate. Teleoconch sutural ramps are flat with 1–3 increasing to 5–7 spiral grooves. Last whorl is with closely spaced, variably granulose spiral ribs. Color of the shell is variable: viz. white, pink, orange, to dark brown. Last whorl is with a white central-band, usually containing brown blotches above and spirally arranged dots below. In light colored specimens, spiral ribs outside the central band occasionally bear dark spiral lines. Shell size varies from 30 to 62 mm. Like all species within the genus Conus, these snails are venomous to humans.

Toxin Type: Conopeptides; venom type not reported

Conus coelinae **(Crosse, 1858)**

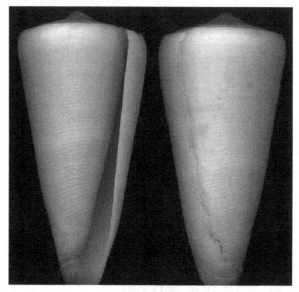

Apertural view   Abapertural view

Common Name: Celine's cone

Geographical Distribution: From the Marshall Islands in the north and the Solomon Islands and Queensland down to New Caledonia; Hawaiian islands; Philippines and Okinawa (Japan) and off the coast of Borneo

Habitat: Intertidal to about 55 m; sandy bottoms.

Identifying Features: This species has a moderately large to large and solid to heavy shell. Last whorl is conical and outline is straight, except convex below shoulder. Shoulder is angulate to sharply angulate. Spire is usually low and outline is slightly sigmoid. Teleoconch sutural ramps are flat to sigmoid, with numerous often-faint spiral striae in later whorls. Entire last whorl is with rather closely spaced spiral threads, usually more prominent basally. Color of the shell is white, variably suffused with yellow. Last whorl is occasionally with a paler spiral band at center. Base is violet or occasionally white. Larval whorls are white. Aperture is white. It feeds on polychaetes. Adults of Conus coelinae can grow to 103 mm but will typically be less than this. Like all species within the genus Conus, these snails are venomous to humans.

Toxin Type: Conopeptides; venom type not reported.

Asprella colmani **(Röckel, D., and W. Korn, 1990) (=** *Conus colmani***)**

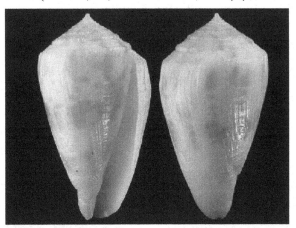

Apertural view   Abapertural view

Common Name: Not designated

Geographical Distribution: Queensland, Australia (Swain Reefs, situated in the southern region of the Great Barrier Reef).

Habitat: Mud and sand between 170 m and 250 m

Identifying Features: Shell of this species is medium-sized and moderately solid. Last whorl is conical or ventricosely conical to broadly conical or broadly and ventricosely conical. Outline is variably convex adapically and straight below. Left side is constricted just above base. Shoulder is angulate. Spire is of low to moderate height and its outline is concave to nearly straight. Teleoconch sutural ramps are flat, with 1 increasing to

5–8 spiral grooves. Last whorl is often with spiral ribs at base and a few weak spiral ribbons above. Last whorl may also have groups of sometimes finely granulose elevations, each consisting of 2–3 fine spiral ribs anteriorly and 1 ribbon or 2 coarse ribs posteriorly. Ground color is white. Last whorl is with 3 axially connected spiral rows of yellowish-brown to orange axial streaks and flames, below shoulder and on each side of center. Adapical markings are partially extending to spire. Aperture is white. Shell size varies from 35 and 52 cm. Like all species within the genus Conus, these snails are venomous to humans.

Toxin Type: Conopeptides; venom type not reported.

Conus crioulus **(Tenorio and Afonso, 2004)**

Apertural view   Axial view   Abapertural view

Common Name: Not designated

Geographical Distribution: Endemic to the Cape Verde Islands where it is found off the north coast of Maio in the area of Praia Real as well as the adjacent bays

Habitat: Found half-buried in sand in crevices and rock platforms, in very shallow water (0.25 to 0.5 m depth).

Identifying Features: This species has a small shell, which is ventricosely conical, with a moderate spire and a rounded shoulder. Outline of the last whorl is with straight or slightly convex sides. Spire is with a straight profile in most cases, with slightly concave, striated sutural ramps and very well marked suture. It has a greenish-brown color with fine equally spaced spiral lines of a darker brown color, interrupted by white dashes and irregular bluish white flecks, tent-shaped in occasions. White flecks and dashes coalesce forming a more dense bluish white spiral band around the middle portion of the last whorl or slightly below. Shoulder and the spire are white, overlaid with greenish-brown blotches often by coalescing fine axial irregular hairlines. Aperture is purplish, with two white bands, one in the middle portion and another one in the upper part. Inner part of the aperture is bluish, and the columella is purple. Adults of the species typically grow to 35 mm in length. These snails are predatory and venomous. As they are capable of "stinging" humans, live ones should not be handled.

Toxin Type: Conopeptides; venom type not reported.

***Conus cuna* (Petuch, 1998) (= *Conus (Purpuriconus) cuna, Aurantius cuna*)**

Abapertural view   Apertural view

Common Name: Not designated

Geographical Distribution: Colombia East; Moro Tupo Island, San Blas Islands, North Panama.

Habitat: At a depth of only 3 m

Identifying Features: Shells of this species are up to 23 mm and are elongated and more slender with high spire. Shell surface is highly sculptured with large, very numerous, closely packed spiral threads, giving the shell a rough texture. This species is having bright salmon-orange early whorls and protoconch. These snails are predatory and venomous. As they are capable of "stinging" humans, live ones should not be handled.

Toxin Type: Conopeptides; venom type not reported.

Conus cuneolus **(Reeve, 1843)**

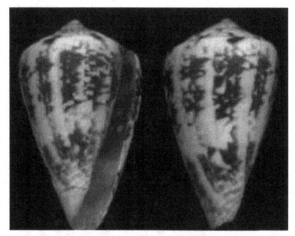

Apertural view   **Abapertural view**

Common Name: Cone snail

Geographical Distribution: West African and Mediterranean regions

Habitat: Either partially buried in fine sandy or gravel bottom under rocks and dead coral slabs, or partially covered with fine muddy sand among small stones and green filament algae.

Identifying Features: This species occurs in two forms. Specimens of form 1 are usually medium to large-sized (between 25 and 35 mm), and are very variable in shell pattern. They are very dark specimens, almost black, or very light, where the netted pattern is reduced to a few black zig-zag dashes forming three spiral bands. Form 2 specimens are with white-colored background and are most often with a bluish shade, contrasting with three well-defined bands of dark brown-netted pattern. Spire tends to be slightly higher than in the first form of this species. These snails are predatory and venomous. As they are capable of "stinging" humans, live ones should not be handled.

Toxin Type: Conopeptides; venom type not reported.

Conus deynzerorum **(Petuch, 1995)** *(= Purpuriconus deynzerorum)*

Apertural view

Common Name: Not designated

Geographical Distribution: Mexico

Habitat: At a depth of only 3 m

Identifying Features: Shell of this species is with a less turbinate shape and is with a much lower, less stepped spire. It has proportionally a very large protoconch, which is composed of two bulbous whorls. Most common color of the shell is bright canary yellow with lighter yellow band central band. Bright pinkish purple specimens do occur. Body whorl is shiny, with 12–15 low, widely spaced spiral cords. Maximum-recorded shell length is 13.5 mm. These snails are predatory and venomous. As they are capable of "stinging" humans, live ones should not be handled.

Toxin Type: Conopeptides; venom type not reported.

Conus dictator **(Melvill, 1898)** **(=** *Conasprella dictator***)**

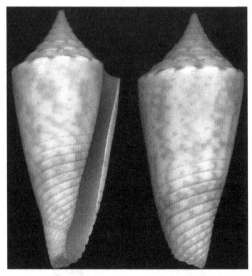

Apertural view   Abapertural view

Common Name: Not designated

Geographical Distribution: From the Persian Gulf, down through the Gulf of Oman and Pakistan, and on to south east India and Sri Lanka; off north Sumatra

Habitat: Depths of around 15–100 m. In shallower waters, this species can be found on coral sand

Identifying Features: Shell of this species is small to medium sized and moderately light. Body whorl is narrowly to broadly conical. Shoulder is angulate. Spire is of moderate height and its outline is concave. Body whorl is with spiral grooves separated by ribbons on basal third to two thirds, sometimes to shoulder. Ground color is cream. Body whorl is with orange to brown axial streaks or flammules, generally fusing in three spiral bands, within the basal third, near center and below the shoulder. Aperture is white marginally and violet within; and its basal portion are often orange. Adults of the species will grow to 48 mm. These snails are predatory and venomous. As they are capable of "stinging" humans, live ones should not be handled.

Conus eldredi (Morrison, 1955)

Abapertural view   **Apertural view**

Common Name: Not designated

Geographical Distribution: Southern Pacific Ocean (Indonesia and French Polynesia).

Habitat: Subtidal but generally not found deeper than 50 m

Identifying Features: Shell is moderately large or moderately light to moderately solid. Last whorl is ovate, narrowly ovate, cylindrical or narrowly cylindrical. Outline varies from moderately convex to almost straight. Left side is concave. Aperture is wider at base than near shoulder. Shoulder is angulate to subangulate and tuberculate. Spire is low and its outline is straight to slightly convex. Postnuclear whorls are tuberculate. Teleoconch sutural ramps are flat and concave in later whorls. Later ramps are with 4 increasing to 10–12 fine spiral grooves. Last whorl is with widely spaced weak spiral ribs at base and shoulder and is with widely spaced, irregularly punctate shallow grooves centrally. Ground color is white, and is suffused with pink and violet. Last whorl is with a fine but usually incomplete network of light brown to dark reddish brown lines and often-triangular spots. Interrupted brown spiral lines and streaks may be present. Apex is white. Later postnuclear sutural ramps are with sparse brown radial lines and streaks on a pink ground. Marginal tubercles are usually white. Aperture is white. Size of the shell varies from 57 to 65 mm. Like all species within the genus Conus, these snails are predatory and venomous. They are known to sting humans and therefore live ones should be handled carefully or not at all.

Toxin Type: Conopeptides; venom type not reported.

Conus erythraeensis **(Reeve, 1843)** **(=*Asprella erythraeensis*)**

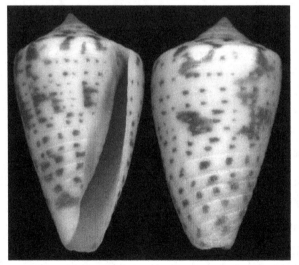

Apertural view   **Abapertural view**

Common Name: Red Sea Cone

Geographical Distribution: Central and S. Red Sea; Aden to Kuria-Muria Is., Oman

Habitat: Shallow water, to about 15 m; very quiet shallow lagoons, in sand and among eel-grass roots

Identifying Features: Shell of this species is small to moderately small or usually moderately light to moderately solid. Last whorl is conical to broadly or ventricosely conical. Outline is convex at adapical fourth to two-thirds and straight below. Left side may be concave near base. Shoulder is angulate. Spire is low to high and its outline is concave to almost straight.

Teleoconch sutural ramps are flat to slightly concave axially, with 1–2 increasing to 3–4 or sometimes 5–6 spiral grooves. Last whorl is with variably wide spiral grooves toward base. Ribbons are seen between narrow or grading to ribs at base. Ground color is white to bluish white. Last whorl is with spiral rows of light reddish or dark brown dots, spots or bars that may fuse into flecks, axial blotches and spiral bands, below shoulder, within adapical and abapical third. Larval whorls are white to brown. Following sutural ramp, it is variably maculated with light to dark brown radial streaks, spots or blotches. Aperture is brown, white, violet or brownish violet, sometimes brown only deep within. Shell size varies from 16 to 35 mm. Like all species within the genus Conus, these snails are predatory and venomous. They are capable of "stinging" humans and therefore live ones should be handled carefully or not at all.

Toxin Type: Conopeptides; venom type not reported.

Conus evorai* (Monteiro, Fernandes, and Rolan, 1995) (=** *Africonus evorai)**

Apertural view   Axial view   Abapertural view

Common Name: Not reported

Geographical Distribution: Endemic to the Cape Verde Islands (Baía das Gatas and the islet of Sal Rei)

Habitat: Rocky reefs in shallow water at typically 0–2 m depth

Identifying Features: Shell is with a relatively convex spire and grows up to 25 mm only. Like all species within the genus Conus, these snails are predatory and venomous. They are capable of "stinging" humans and therefore live ones should be handled carefully or not at all.

Toxin Type: Conopeptides; venom type not reported.

Conus eximius (Reeve, 1849)

Abapertural view   Apertural view

Common Name: The Choice Cone

Geographical Distribution: Throughout Papua New Guinea, the Philippines and Taiwan, extending to the Bay of Bengal and surrounding areas; Western Australia

Habitat: At depths between 20 and 100 m on sandy substrate.

Identifying Features: Shell of this species is medium sized and is light to solid. Body whorl is conical. Outline is slightly convex near shoulder and straight below. Shoulder is angulate. Spire is low to high and it s outline is concave to deeply concave. Basal half of body whorl is with variably spaced spiral grooves separating ribs anteriorly and a few ribbons posteriorly. Outer lip is of aperture thick and nearly straight. Ground color is white. Body whorl is with broad, and mostly interrupted brown spiral bands are seen on either side of the center. Brown axial flames extend from the posterior brown band to shoulder. Irregularly spaced band may also be seen below shoulder. Aperture is white. Maximum size for shells of this species is 58 mm in length. Like all species within the genus Conus, these snails are predatory and venomous. They are capable of "stinging" humans and therefore live ones should be handled carefully or not at all.

Toxin Type: Conopeptides; venom type not reported.

Pyruconus fergusoni (Sowerby, G.B. III, 1873) (= *Conus fergusoni*)

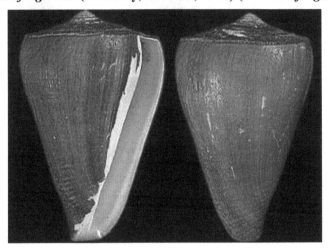

Apertural view   **Abapertural view**

Common Name: Ferguson's cone

Geographical Distribution: Pacific Ocean along the Galapagos Islands and the from the Sea of Cortez to Peru

Habitat: Subtidal sand, patch reefs and rubble at depths between 5 and 150 m deep.

Identifying Features: This species has a heavy shell, which has a slightly channeled spire. The white shell is lightly striolate transversely and is covered under a brown epidermis. Size of an adult shell varies between 60 mm and 150 mm. Like all species within the genus Conus, these snails are predatory and venomous. They are capable of "stinging" humans and therefore live ones should be handled carefully or not at all.

Toxin Type: Conopeptides; venom type not reported.

Conus fijiensis **(Moolenbeek, Röckel, and Bouchet, 2008)**

Abapertural view

Common Name: Not designated

Geographical Distribution: Fiji, South-East of Viti Levu

Habitat: Not reported

Identifying Features: Shell of this species is medium to small size, thin, narrowly conical and slightly pyriform. Spire is slightly concave with slightly stepped whorls. Protoconch is blunt and paucispiral with 2 smooth, transparent whorls. Teleoconch is of 7.7 whorls with nodular shoulder. First whorl is with 12 sharp nodules, which are gradually fading out in subadult and adult whorls. Shoulder ramp is with 3 or 4 fine spiral threads and numerous fine axial wrinkles. Last whorl (with one large repaired scar) is with 32 grooves. Color of last whorl is white with two ill-defined brown bands, an irregular pattern of axially elongated brown blotches, and one fine, semicontinuous, brown line in the center of many groove interspaces. Spire is white with regular and radiating brown patches. There are 5 such patches on last whorl and a little darker patch is near shoulder. Aperture is white. Height and width of the shell are 17.8 mm 6.8 mm, respectively.

Like all species within the genus Conus, these snails are predatory and venomous. They are capable of "stinging" humans and therefore live ones should be handled carefully or not at all.

Toxin Type: Conopeptides; venom type not reported

Conus fijisulcatus **(Moolenbeek, Röckel, and Bouchet, 2008)** *(= Asprella fijisulcata)*

Apertural view   Abapertural view

Common Name: Not reported as it is a new species

Geographical Distribution: Fiji

Habitat: Not reported

Identifying Features: Shell of this species is medium to large, broadly conical and moderately solid. Spire is rather high and straight. Protoconch is of 2.3 whorls, smooth, glossy and transparent. Teleoconch is of 11.2 whorls. Whorls are slightly concave with 2 spiral grooves in the first teleoconch whorl and are gradually growing to about 5 grooves axially crossed by opisthocline growth markings. First teleoconch whorl is with about 14 fine nodules, which are gradually diminishing and lacking on the last two whorls. Suture is deep. Color on the first whorls is dominated by brown blotches and is gradually becoming more white. Shoulder is angulate with a prominent white rim. Body whorl is straight and slightly convex near the shoulder, with 3–4 repair scars. Upper part is smooth and other 3/4 is with about 30 spiral groves. In these grooves fine axial riblets are seen. On the lower spiral ribs fine nodules are seen. Color is brown with a few white spirals of which the one in the middle is most prominent. Base is white. Shell size varies from 18 to 55 mm. Like all species within the genus Conus, these snails are predatory and venomous. They are capable of "stinging" humans and therefore live ones should be handled carefully or not at all.

Toxin Type: Conopeptides; venom type not reported.

Conus fontonae (Rolán & Trovão in Rolán, 1990)

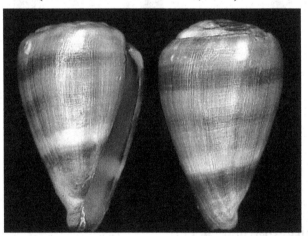

Apertural view   **Abapertural view**

Common Name: Not designated

Geographical Distribution: Endemic to the Cape Verde Islands where it has been found only off the north-western coast of the island of Sal.

Habitat: Bays of calm waters; large boulders and rock platforms

Identifying Features: This species is characterized by the presence of several spiral bands, usually well-defined, being of different shades of greenish-brown. These bands appear always below the shoulder and around the base, and most often also in the midbody of the last whorl. Some specimens display very fine axial lines of greenish brown color. Shell size varies from 20 to 25 mm only. Like all species within the genus Conus, these snails are predatory and venomous. They are capable of "stinging" humans and therefore live ones should be handled carefully or not at all.

Toxin Type: Conopeptides; venom type not reported.

Conus frigidus **(Reeve, 1848) (=** *Conus (Virgiconus) frigidus***)**

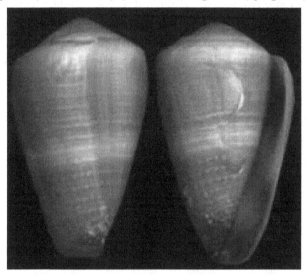

Abapertural view   Apertural view

Common Name: Frigid cone

Geographical Distribution: Tropical Central and W. Pacific except for Hawaii.

Habitat: On intertidal benches and shallow subtidal reef flats to about 5 m; bare beachrock or limestone, beachrock and limestone pavement with a thin layer of sand or with algal turf, sand-filled depressions, coral rubble with or without sand, and dead coral heads or rocks.

Identifying Features: Shell of this species is moderately small to medium-sized or moderately solid-to-solid. Last whorl is conical or ventricosely conical to broadly conical. Outline is convex at adapical third to half, almost straight below. Shoulder is angulate, sometimes subangulate or nearly carinate. Spire is of low to moderate height and its outline is straight to convex. Teleoconch sutural ramps are usually flat. Later ramps are with 2–4 increasing to 4–5 distinct spiral grooves and additional striae. Last whorl is usually with variably spaced, generally granulose spiral ribs from base to center or shoulder. Surface is seldom largely smooth. Color is olive to sometimes orangish or pinkish brown. Last whorl is usually with a paler spiral band at or closely below center and sometimes a second one at shoulder. Base is purplish blue. Larval shell is purple. Aperture is purple, with paler bands near center and below shoulder, often grading to pale blue deep within. Shell size varies from 28 to 56 mm. Like all species within the genus Conus, these snails are predatory and venomous. They are capable of "stinging" humans and therefore live ones should be handled carefully or not at all.

Toxin Type: Conopeptides; venom type not reported.

Conus fragilissimus (Petuch, 1979)

Apertural view   Abapertural view

Common Name: Fragile Geography cone

Geographical Distribution: Central and S. Red Sea

Habitat: At depths of 2 to 5 m in mud or coral rubble on coral reefs

Identifying Features: Shell of this species is moderately small to medium-sized or light to moderately light. Last whorl is ovate. Outline is convex and left side is slightly concave at basal third. Aperture is wider at base than near shoulder. Shoulder is angulate and tuberculate. Spire is of low to moderate height and its outline is concave. Larval shell is of 1.75 whorls with a maximum diameter about 0.9–1 mm. Postnuclear whorls are tuberculate and the tubercles are weak in early whorls. Teleoconch sutural ramps are flat to slightly concave, with 0–1 increasing to 4–6 weak or pronounced spiral grooves. Last whorl is with a few weak, widely spaced spiral ribs at base. Ground color is white to gray. Last whorl is with a fine, incomplete, light brown to dark reddish brown reticulate pattern fusing in variously sized and shaped blotches that leave an interrupted spiral ground-color band at center. Spiral rows are of brown dots and dashes, often alternating with white dots and dashes, extend from base to shoulder but may be weak. Larval whorls and adjacent 2 postnuclear sutural ramps are dark brown. Following sutural ramps, dark brown radial streaks and blotches are seen. Aperture is translucent, shaded with violet. Shell size varies from 26 to 47 mm. The species is piscivorous. Like all species within the genus Conus, these snails are predatory and venomous. They are capable of "stinging" humans and therefore live ones should be handled carefully or not at all.

Toxin Type: Conopeptides; venom type not reported.

Conus frausseni (**Tenorio and Poppe, 2004**) (= *Profundiconus frausseni*)

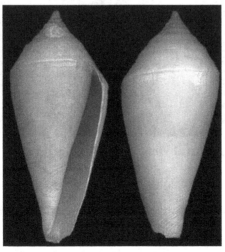

Apertural view   Abapertural view

Common Name: Fraussen's cone

Geographical Distribution: Aliguay Island, Philippines

Habitat: Not reported

Identifying Features: This species has medium sized shell and the profile of the shell is conical or ventricosely conical with a moderate spire and a rounded shoulder. Outline of last whorl is straight or sigmoid and slightly pyriform. Outline of spire is rather sigmoid with a prominent protoconch. Ground color of the shell is orange brown and color of aperture is orange. Shell size varies from 26 to 47 mm. Like all species within the genus Conus, these snails are predatory and venomous. They are capable of "stinging" humans and therefore live ones should be handled carefully or not at all.

Toxin Type: Conopeptides; venom type not reported.

Conus fumigatus **(Hwass in Bruguière, 1792)** *(= Conus luctificus)*

Apertural view   Abapertural view

Common Name: Smoky cone

Geographical Distribution: Mediterranean and Red Sea

Habitat: Shallow waters

Identifying Features: Shell of this species is medium sized and thick, with regularly conical, shouldered body whorl and very low conical spire. Sculpture is limited to faint spiral grooves at the extremity of the body whorl, towards the siphonal canal and spiral striae between the shoulder and the suture. Early whorls of spire are also tuberculate. Aperture is elongated, narrow and parallel-sided, with thin outer lip. Color of the shell is mostly brownish to greenish with superimposed darker interrupted spiral lines on the body whorl. A whitish zone is seen along the suture and another one on mid-body whorl, marked with irregular axial flames. Inside of aperture is grayish. Shell size varies from 32 to 60 mm. Like all species within the genus Conus, these snails are predatory and venomous. They are capable of "stinging" humans and therefore live ones should be handled carefully or not at all.

Toxin Type: Conopeptides; venom type not reported.

Conus genuanus (Linnaeus, 1758) (= *Genuanoconus genuanus*)

Apertural view   Abapertural view

Common Name: Garter cone

Geographical Distribution: Atlantic Ocean from the Canary Islands and Cape Verde to Angola; in the Indian Ocean along Madagascar

Habitat: Mud and sand at depths ranging from 1 m down to 20 m

Identifying Features: Ground color of the shell is pink-brown or violaceous brown, with revolving narrow lines of alternate white and chocolate quadrangular spots and dashes. These lines are usually alternately larger and smaller. Surface of the shell is usually smooth, but sometimes the lines are slightly elevated. Spire is smooth. Shell size varies from 33 to 75 mm. Like all species within the genus Conus, these snails are predatory and venomous. They are capable of "stinging" humans and therefore live ones should be handled carefully or not at all.

Toxin Type: Conopeptides; venom type not reported.

Conus gigasulcata (Moolenbeek, Röckel, and Bouchet, 2008)

Apertural view   **Abapertural view**

Common Name: Sea snail

Geographical Distribution: Pacific Ocean along Fiji and Vanuatu

Habitat: At depths of between 150 and 183 m on soft bottoms

Identifying Features: Shell of this species is large, broadly conical and moderately solid. Spire is rather low and slightly concave. Teleoconch is of about 11 whorls. Whorls are slightly concave with 2 spiral grooves in the first teleoconch whorl. Last whorl is with 12 strong nodules. Suture is rather deep. Color is basically white and is interrupted by brown blotches. Shoulder is angulate. Last whorl is straight, and is slightly convex near the shoulder. Adapical half is smooth and abapical part is with about 26 spiral grooves, crossed by irregular fine axial riblets. On the abapical spiral cords, a few small nodules are seen. Color is dark white with a pattern of brown dotted spirals. In the middle these brown/white spiral lines have a broader band of brown blotches. Base is white. Periostracum is rather thick and brown to dark brown. Size of an adult shell varies between 28 mm and 90 mm. Like all species within the genus Conus, these snails are predatory and venomous. They are capable of "stinging" humans and therefore live ones should be handled carefully or not at all.

Toxin Type: Conopeptides; venom type not reported.

Conus glans **(Dautzenberg, 1937)** **(= *Leporiconus glans*)**

Abapertural view and vice-versa   Abapertural view

Common Name: Acorn cone

Geographical Distribution: Throughout the Indo-Pacific excluding Hawaii

Habitat: Subtidal habitats in particular on coral reef slopes, on coarse sand and under coral reef.

Identifying Features: Shell of this species is medium sized and elongately cylindrical. Spire is convex and apex is obtuse. Shell is transversely striated and striations are finely granulated or smooth. Granules are found compressed. Shell has well-developed siphonal canal which is elongated and trunk like. Color is violet or yellowish brown with intermittent bands of white. Once mature, it can reach a size ranging from 20 mm to 60 mm These snails are predatory and venomous. They are capable of "stinging" humans and therefore live ones should be handled carefully or not at all.

Toxin Type: Conopeptides; venom type not reported.

Conus gordyi **(Röckel and Bondarev, 1999)**

Apertural view   Abapertural view

Common Name: Not reported

Geographical Distribution: Endemic to the Saya de Malha Bank, located between the Seychelles, Mauritius and Chagos

Habitat: This species has been recently described and there is little information about this cone snail's ecology. It is presumed to inhabit in sandy silt and limestone debris at depths down to 130 m.

Identifying Features: In this species, postnuclear whorls, except for last whorl are strongly tuberculate. Last whorl is with about 25 spiral ribbons from base to shoulder, separated by narrow spiral grooves with close-set axial threads. Ground color is white and last whorl is with 3–4 bands of orange bars or rectangular dashes. Spire is variably spotted with the same color. Aperture is matching the exterior coloration, at times slightly violet basally.

Shell is up to 19.8 mm. Like all species within the genus Conasprella, these snails are predatory and venomous. They are capable of "stinging" humans and therefore live ones should be handled carefully or not at all.

Toxin Type: Conopeptides; venom type not reported.

Conus grangeri **(G. B. Sowerby III, 1900) (=** *Phasmoconus grangeri***)**

Apertural view   Abapertural view

Common Name: Granger's cone

Geographical Distribution: Southern Red Sea, Philippines and off Sri Lanka

Habitat: Silt or sand at depths of around 210–800 m

Identifying Features: Shell of this species is moderately small and its profile is broadly conical with a moderate soiree and a carinate shoulder. Outline of last whorl is sigmoid convex adapically and concave basically. Shape is pyriform. Spire is concave with a concave sutural ramp. Protoconch is with 3 globose white and translucent whorls. Three bands of interrupted cloudy brown blotches are present. Shell size ranges from 31 to 75 mm. Like all species within the genus Conus, these snails are predatory and venomous. They are capable of "stinging" humans and therefore live ones should be handled carefully or not at all.

Toxin Type: Conopeptides; venom type not reported.

Conus gubernator (Hwass, 1792)

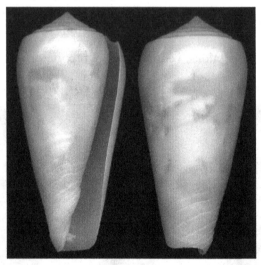

Apertural view   **Abapertural view**

Common Name: Governor cone

Geographical Distribution: Western and central Indian Ocean from Natal, South Africa and Madagascar; southern India and Sri Lanka

Habitat: Shallow and deeper water to 60 m; sand, coral rubble and sandy grass.

Identifying Features: Shell of this species is moderately large; solid to moderately heavy and glossy. Body whorl is ventricosely conical. Outline is slightly convex adapically and straight below. Shoulder is angulate. Spire is of moderate height and bluntly pointed. Its outline is slightly convex. Body whorl is with several shallow spiral grooves on basal fourth to third and rest of the whorl is smooth. Aperture is uniformly wide. Outer lip is thin, sharp and straight. Ground color is white and is suffused with light brown. Body whorl is with separated blackish-brown axial markings. These markings are variable in shape and their size is ranging from irregular flecks to large, zig-zag flames. Early spire whorls are pinkish. Aperture is white. This is a fish hunting species that may grow to 106 mm but will generally be less than this. Like all species within the genus Conus, these snails are predatory and venomous. They are capable of "stinging" humans and therefore live ones should be handled carefully or not at all.

Toxin Type: Conopeptides; venom type not reported.

Conasprella guidopoppei **(Raybaudi Massilia, 2005)** (= *Conus guidopoppei*)

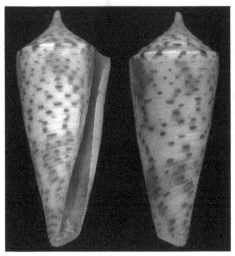

Apertural view   Abapertural view

Common Name: Not designated

Geographical Distribution: South of the Philippines and in Sabah

Habitat: At depths between 10 and 30 m

Identifying Features: Shell of this species is creamy white to pink to orange with irregular brown markings in the form of tiny dots or spots. Once mature, it can reach a size ranging from 20 mm to about 38 mm. Like all species within the genus Conus, these snails are predatory and venomous. They are capable of "stinging" humans and therefore live ones should be handled carefully or not at all.

Toxin Type: Conopeptides; venom type not reported.

Tenorioconus hanshassi (Lorenz and Barbier, 2012)

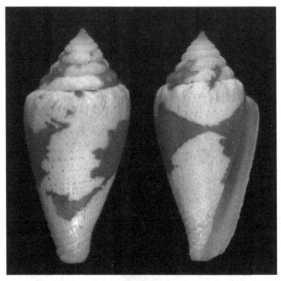

Abapertural view   Apertural view

Common Name: It is a new species

Geographical Distribution: Type locality is Siargao Island, Philippines

Habitat: Shallow water of 20 m depth

Identifying Features: Shell of this species is generally elongated conical to turgid in shape, with an elevated spire and a definite shoulder. It has a white base color and is marked with medium brown splotches and irregular rows of dots or dashes. The holotype is 23.3 mm in length. Like all species within the genus Conus, these snails are predatory and venomous. They are capable of "stinging" humans and therefore live ones should be handled carefully or not at all.

Toxin Type: Conopeptides; venom type not reported.

Conus helgae **(Blöcher, 1992)** (= *Rolaniconus helgae*)

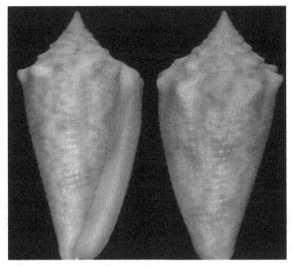

Apertural view   Abapertural view

Common Name: Not designated

Geographical Distribution: Endemic to Madagascar; southern coast extending up to the north west; central Madagascan coast.

Habitat: Deep water species from 75 m to 600 m. There are no other data in the literature on the species habitat or ecology.

Identifying Features: Shell of this species is medium-sized and moderately solid. Last whorl is ventricosely conical-to-conical. Outline is convex at right side and sigmoid to nearly straight at left side. Shoulder is angulate and strongly tuberculate. Tubercles are widely spaced and outward-pointing. Spire is of moderate height and its outline is concave. Larval shell is of about 2 whorls with a maximum diameter about 1 mm. Postnuclear spire whorls are tuberculate. Teleoconch sutural ramps are concave, with 1 increasing to 4–5 spiral grooves; grooves with a spiral thread on last ramp. Last whorl is with granulose spiral ribs from base to subshoulder area, followed by weaker ribs; a spiral thread often present anterior to the ribs. Ground color is white. Last whorl is with spirally aligned brown bars and flecks, the latter concentrated in a spiral band on each side of center. Base and basal part of columella are pinkish violet. Larval shell is white.

Postnuclear sutural ramps are with brown radial streaks. Aperture is tinged with pinkish violet. Size of the shell ranges from 36 to 37 mm. Like all species within the genus Conus, these snails are predatory and venomous. They are capable of "stinging" humans and therefore live ones should be handled carefully or not at all.

Toxin Type: Conopeptides; venom type not reported.

Conus hirasei **(Kuroda, 1956)** (= *Kioconus hirasei*)

Apertural view   **Abapertural view**

Common Name: Hirases cone

Geographical Distribution: Japan to Philippines and South China Sea

Habitat: At depths of 100–240 m

Identifying Features: Shell of this species is medium-sized to large and solid. Last whorl is conical, ventricosely conical or slightly pyriform. Outline is convex adapically and straight to slightly concave below. Shoulder is angulate to subangulate and is slightly carinate. Spire is of low to moderate height and its outline is slightly concave to slightly sigmoid. Early postnuclear whorls are tuberculate. Teleoconch sutural ramps are flat, with 3 increasing to 5–6 spiral grooves. Additional spiral threads are seen on last 2 ramps. Last whorl is with a few broad but rather weak spiral ribs at base. Ground color is white tinged with violet. Last whorl is with about 18–30 rather regularly spaced reddish brown spiral lines from base to subshoulder area. Larval whorls are white. Outer margins of teleoconch sutural ramps are with regularly spaced reddish to blackish brown spots. Aperture is white except for violet area near shoulder. This is a vermivorous species preying on polychaete worms. Size of the shell ranges from 50 to 92 mm. Like all species within the genus Conus, these snails are predatory and venomous. They are capable of "stinging" humans and therefore live ones should be handled carefully or not at all.

Toxin Type: Conopeptides; venom type not reported.

Conus hyaena (Hwass in Bruguière, 1792)

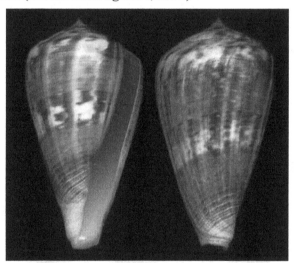

Apertural view   Abapertural view

Common Name: Hyena cone

Geographical Distribution: Indian Ocean (Madagascar and Bay of Bengal); Pacific Ocean (Philippines and Indonesia); South China Sea.

Habitat: Intertidal and slightly subtidal; from fine sand to coarse gravel or rock

Identifying Features: Shell of this species is medium to moderately large and moderately solid-to-solid. Body whorl is conical. Uppersides are convex near shoulder and less so or conical below. Left side is concave near base. Shoulder is angulate to rounded. Spire is of low to moderate height and its outline is straight to slightly convex. Aperture is uniformly wide. Outer lip is sharp and evenly convex. Body whorl is dull white to yellow and is heavily covered with broad reddish brown to dark brown streaks. Usually a paler mid-body area is visible through streaks. Spire and shoulder are dirty white to brown and are heavily covered with curved light brown lines and blotches, about the same color as the body whorl. Aperture is bluish white. Size of an adult shell varies between 29 mm and 80.5 mm. These snails are predatory and venomous. They are capable of "stinging" humans and therefore live ones should be handled carefully or not at all.

Toxin Type: Conopeptides; venom type not reported.

Conus immelmani **(Korn, 1998)** *(= Nataliconus immelmani)*

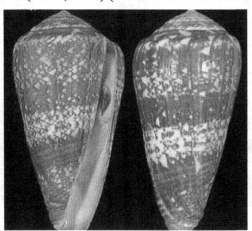

Apertural view   Abapertural view

Common Name: It is a new species and is not designated with common name

Geographical Distribution: Endemic to South Africa

Habitat: There is little information on habitats of this species in the literature. It is referred to as being caught on coral reef; depth range is 35–50 m

Identifying Features: Shell of this species is white in adults and lilac in subadult. Early postnuclear whorls are also lilac. Last whorl is with 3 spiral bands with reticulated pattern of olive-tan to brown. Spiral rows of dark brown dots to short fine axial dashes extend from base to shoulder. Shell size is up to 90 mm. It is a mollusciverous species. Like all species within the genus Conus, these snails are predatory and venomous. They are capable of "stinging" humans and therefore live ones should be handled carefully or not at all.

Toxin Type: Conopeptides; venom type not reported.

Conus infrenatus **(Reeve, 1848) (= *Sciteconus infrenatus*)**

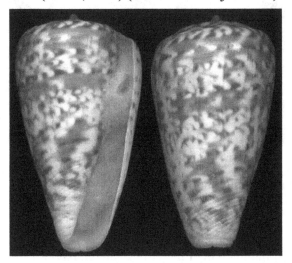

Apertural view   Abapertural view

Common Name: Jeffrey's Bay cone

Geographical Distribution: Endemic to South Africa; off the Transkei Coast, from Jeffrey's Bay to Coffee Bay, and off the coast of East London

Habitat: There is little information for this species' habitats; coral reefs at depths of 15–60 m

Identifying Features: Size of an adult shell varies between 24 mm and 50 mm. Shell is rosy white, and is encircled by articulated lines of chestnut and white spots. Apex is pink. Like all species within the genus Conus, these snails are predatory and venomous. They are capable of "stinging" humans and therefore live ones should be handled carefully or not at all.

Toxin Type: Conopeptides; venom type not reported.

Conus iodostoma **(Reeve, 1843)**

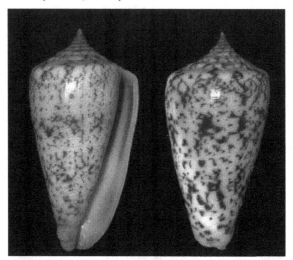

Apertural view   Abapertural view

Common Name: Violet-mouth cone

Geographical Distribution: Indian Ocean along Madagascar and Mozambique

Habitat: Intertidal to 20 m or more; in sheltered bays on soft substratum

Identifying Features: Shell of this species is moderately small to medium-sized, and moderately light to moderately solid. Last whorl is conical or ventricosely conical. Outline is convex at adapical third and less so or straight below. Left side is often slightly concave above base. Shoulder is subangulate. Spire is of moderate height and its outline is concave. Larval shell is of 2–2.25 whorls with a maximum diameter 0.8–1 mm. Teleoconch sutural ramps are flat, with 2 increasing to 4–5 spiral grooves. Last whorl is with rather widely spaced punctate spiral grooves from base to center or slightly beyond. Grooves are wider and interstitial ribbons are grading to ribs at anterior end. Ground color is white to bluish gray. Last whorl is with few to numerous spiral rows of brown to reddish-brown dots; sometimes with wavy axial lines forming an irregular network, often concentrated on each side of center and also below shoulder. Axial lines may concentrate in solid or interrupted spiral bands. Shell size ranges from 28 to 42 mm. Like all species within the genus Conus, these snails are predatory and venomous. They are capable of "stinging" humans and therefore live ones should be handled carefully or not at all.

Toxin Type: Conopeptides; venom type not reported.

Conus isabelarum (Tenorio and Afonso, 2004)

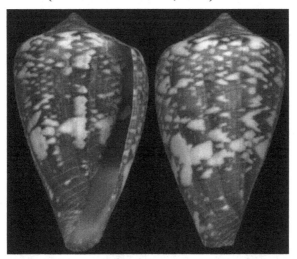

Apertural view   Abapertural view

Common Name: Not designated

Geographical Distribution: Baía de Navío Quebrado

Habitat: At 2–4 m depth; attached to the bottom of rocks and dead coral slabs on sandy bottom, or partially buried in sand under rock.

Identifying Features: Shell of this species is small to moderately small, and ventricosely conical, with a low to moderate spire and a rounded shoulder. Outline of the last whorl is rather convex. Concave spire is striated, with flat to slightly convex sutural ramps. Prominent protoconch is measuring ca. 0.7 mm. Shell has a rich honey-brown color, with fine spiral lines of darker brown very often visible. Ground color is overlaid with irregular white markings, tent-shaped in many cases. White marks are especially evident in a spiral band slightly below the midbody of the shell. Shoulder and the spire are patterned with irregular brown and white blotches. Aperture is white, but a purplish diffuse blotch is often present in the upper part, especially in the smaller specimens. Shell length varies from 17 to 30 mm. Like all species within the genus Conus, these snails are predatory and venomous. They are capable of "stinging" humans and therefore live ones should be handled carefully or not at all.

Toxin Type: Conopeptides; venom type not reported.

Conus (Magelliconus) jacarusoi **(Petuch, 1998)**

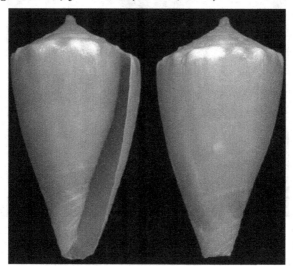

Apertural view   **Abapertural view**

Common Name: Not designated

Geographical Distribution: Gulf Mexico E. Colombia

Habitat: At a depth of 15 m

Identifying Features: Shell of this species is elongated and conical in shape. Spire is well elevated and scalariform. Shoulders are also carinate at all growth stages. Outline of the last whorl is elongate conical with straight to just slightly convex sides. Series of spiral ribbons are present. Aperture is white colored inside. Coloration consists of dark to light brown markings. These are mostly longitudinal in orientation but traces of spiral markings are present when spiral ribbons are well developed. Basic pattern consists of three zones of color markings. Ground color is white. Maximum shell length is 25 mm. Like all species within the genus Conus, these snails are predatory and venomous. They are capable of "stinging" humans and therefore live ones should be handled carefully or not at all.

Toxin Type: Conopeptides; venom type not reported.

Conus janus **(Hwass in Bruguiere, 1792)**

Apertural view   Abapertural view

Common Name: Not designated

Geographical Distribution: E. C. Africa; Mascarenes; S. India; Philippines

Habitat: In 8–20 m; on sand and fine shell debris with sparse sea-weed, exposed to strong tidal currents.

Identifying Features: Shell of this species is medium-sized to moderately large and moderately solid-to-solid. Last whorl is conical and occasionally ventricosely conical. Outline is convex adapically and less so to straight below. Left side may be slightly concave at base. Shoulder is subangulate. Spire is of low to moderate height and its outline is deeply concave. Larval shell is of about 3 whorls with a maximum diameter about 0.7 mm.

Teleoconchsutural ramps are flat to slightly concave, with 2 increasing to 4–6 spiral grooves. Last whorl is with a few widely spaced, axially striate grooves around basal fourth. Ribbons are divided into several fine ribs towards anterior end. Ground color is white. Last whorl is with 3 variably broad yellowish to dark brown or orange spiral bands, below shoulder and above and well below center. Bands are occasionally with darker brown spiral lines or variably reduced. Straight or wavy, yellowish to dark brown axial streaks of variable width extend from base to shoulder and onto spire. Streaks are separate or confluent and continuous or interrupted. Base is pale yellow. Apex is brown. Later sutural ramps are with brown to blackish brown radial markings. Aperture is white or suffused with pale yellow or orange. Shell length ranges from 45 to 75 mm. Like all species within the genus Conus, these snails are predatory and venomous. They are capable of "stinging" humans and therefore live ones should be handled carefully or not at all.

Toxin Type: Conopeptides; venom type not reported.

Conus jaspideus **(Gmelin, 1791)**

Abapertural view　　　Apertural view

Common Name: Not designated

Geographical Distribution: Continental shelf area of the northern coast of South America with records from Colombia (off Guajira peninsula), Venezuela, Trinidad and Brazil

Habitat: Muddy sand or silt at depths of 20–120 m.

Identifying Features: This species has a moderately heavy shell of 20 to 30 mm size. Shell is biconical, with straight to slightly concave-sided, rather high, stepped spire (at least 1/3 of total length) and straight-sided body whorl. Shoulder is angulate and body whorl is with heavy granules placed on broad spiral ridges between incised lines. In some specimens also granules on the shoulder and on the margin of some earlier whorls. Shoulder may also be somewhat undulate. Tops of the whorls are rather flat without spiral grooves and with only thin curved axial growth lines. Spire whorls are with carinate margins. Color is white to very pale light brown with reddish brown clouds. Granules are whitish and quite distinct against a reddish brown background. In some specimens small brown dashes can be distinguished between the white granules. Spire is white with axial flammules. Like all species within the genus Conus, these snails are predatory and venomous. They are capable of "stinging" humans and therefore live ones should be handled carefully or not at all.

Toxin Type: Conopeptides; venom type not reported.

Conus jeanmartini (Raybaudi G. Massilia, 1992)

Apertural view   Abapertural view

Common Name: Not designated

Geographical Distribution: Reunion

Habitat: At depths greater than 500 m

Identifying Features: Shell of this species is medium-sized and moderately light. Last whorl is conical to ventricosely conical. Outline is slightly convex below shoulder and straight toward base. Shoulder is subangulate to rounded, with a prominent edge. Spire is of moderate height and its outline is slightly sigmoid. Maximum diameter of larval shell is about 1 mm. First 4–7 postnuclear whorls are tuberculate. Teleoconchsutural ramps are slightly concave, with obsolete spiral striae. Last whorl is with fine spiral ribs at base and spiral threads above. Color is beige, with an inconspicuous narrow light brown spiral band just below shoulder edge. Aperture is cream. Shell length is up to 42 mm. Like all species within the genus Conus, these snails are predatory and venomous. They are capable of "stinging" humans and therefore live ones should be handled carefully or not at all.

Toxin Type: Conopeptides; venom type not reported.

Conus jesusramirezi **(Cossignani, 2010)**

Abapertural view   **Apertural view**

Common Name: Not designated

Geographical Distribution: Guajira Peninsula, E. Colombia

Habitat: No Data

Identifying Features: Shell of this species is of small to average size for the genus. High spire is slightly concave and turreted. Protoconch is of 1.5 whorls. A total of 14 whitish cream tubercles are present on shoulder. Body whorl is elongate and aperture is about 60% of total height. Body whorl is with spiral rows of small flesh colored granules. Base color is beige/cream brown. Size of the shell is 32 × 13 mm. Like all species within the genus Conus, these snails are predatory and venomous. They are capable of "stinging" humans and therefore live ones should be handled carefully or not at all.

Toxin Type: Conopeptides; venom type not reported.

Conus jickelii **(Weinkauff, 1873 (=** *Phasmoconus jickelii***))**

Apertural view   **Abapertural view**

Common Name: Jickeli's cone

Geographical Distribution: S. Red Sea and Gulf of Aden.

Habitat: In 1–25 m

Identifying Features: Shell of this species is medium-sized and moderately solid. Last whorl is conical to ventricosely conical. Outline is slightly to moderately convex at adapical fourth or third, usually straight below. Aperture is somewhat wider at base than near shoulder. Shoulder is angulate to subangulate. Spire is of low to moderate height and its outline is concave. Larval shell is of 2–2.25 whorls with a maximum diameter 0.7–0.8 mm. Teleoconch sutural ramps are flat to slightly concave adaxially, with 1 increasing to 3–4 major spiral grooves, containing spiral striae and threads that may produce 5–7 unequal grooves in last 2 whorls. Last whorl is with variably wide spiral grooves near base, separating ribs anteriorly and a few ribbons posteriorly. Ground color is white to bluish gray. Entire last whorl is with spiral rows of dark reddish or bluish brown dots, dashes, squarish spots and bars that fuse into flames and irregular blotches below shoulder and within adapical as well as abapical third. Larval whorls are gray to brown. About 2 adjacent postnuclear sutural ramps are brown. Late ramps are with radial streaks. Aperture is almost white to pale blue deep within, often with a yellow or brownish violet collabral band behind the translucent marginal zone. Size of the shell ranges from 35 to 51 mm. Like all species within the genus Conus, these snails are predatory and venomous. They are capable of "stinging" humans and therefore live ones should be handled carefully or not at all.

Toxin Type: Conopeptides; venom type not reported.

Conus joliveti (Moolenbeek, Röckel, and Bouchet, 2008)

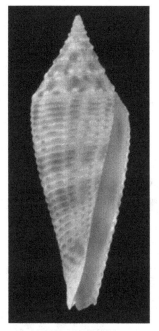

Apertural view

Common Name: Not designated

Geographical Distribution: Fiji

Habitat: At depths of 250–500 m

Identifying Features: Shell of this species is medium-sized for the genus and is thin. Shape is narrowly conical and slightly pyriform. Protoconch is multispiral, consisting of nearly 3 glossy, convex whorls. Protoconch whorls are smooth except for 5 curved axial ribs just before the proto-conch/teleoconch transition. Teleoconch is of 8 whorls and spire is slightly concave. Suture is deep and impressed. Spire whorls are with strongly tubercular keel situated at periphery on first three whorls, then on subsequent whorls gradually less pronounced and situated lower on the whorl, with a proportionally broader ramp. 18 sharp tubercles are present on first teleoconch whorl. Last whorl is with 33 smooth, convex spiral cords. Color of the protoconch is creamy semitransparent. Overall teleoconch color is creamy white with 3 rather distinctly set off, brown, broad spiral bands, and less well defined, narrow, axial stripes alternatingly brown and white. Spire is creamy white with irregular axial brown patches extending from suture to shoulder. Shoulder is with spirally arranged white and brown streaks that do not correspond regularly with the tubercles. Height of the shell is 29.1 mm, and width 10.1 mm. Like all species within the genus Conus, these snails are predatory and venomous. They are capable of "stinging" humans and therefore live ones should be handled carefully or not at all.

Toxin Type: Conopeptides; venom type not reported.

Conus jorioi **(Petuch, 2013)**

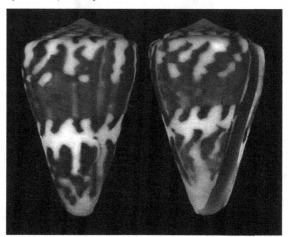

Abapertural view　　**Apertural view**

Common Name: Not reported

Geographical Distribution: Brazil

Habitat: Pen rock platform bottom, occupied by a hermit crab, in 5 m depth

Identifying Features: Shell of this species is heavy and thickened, broadly conical shell and with straight sides. Spire is slightly elevated, broadly subpyramidal and distinctly stepped. Shoulder is sharply angled and edged by thin carina. Body whorl is smooth and silky (slightly pitted and eroded on the holotype), with 3 faintly raised spiral cords around anterior tip.

Shell base color is white, overlaid with 2 broad bands of reddish-brown and orange-brown amorphous flammules, one around posterior two thirds of shell and one around anterior one third of shell; broad brown bands separated by wide white band edged by dark brown, widely separated tooth-shaped flammules. Edge of shoulder and posterior one third of shell are marked with intermittent, evenly separated large white longitudinal flammules. Spire is white with scattered small dark reddish-brown crescent-shaped flammul. Anterior tip of shell is salmon-orange. Aperture is proportionally very narrow and pale salmon on interior. Like all species within the genus Conus, these snails are predatory and venomous. They are capable of "stinging" humans and therefore live ones should be handled carefully or not at all.

Toxin Type: Conopeptides; venom type not reported.

Conus josephinae **(Rolán, 1980)**

Aperture view   Axial view   Abapertural view

Common Name: Not reported

Geographical Distribution: MaioBoavista, Cape Verde Islands

Habitat: Shallow water in about 1–5 m.

Identifying Features: This species has a small (normal length: 25 to 28 mm), solid shell, with a nearly straight profile and a rounded shoulder. Spire is low, with a slightly concave profile. Sutural ramps are with a couple of spiral grooves in younger specimens, which tend to disappear in older ones. Shell is uniformly brown, light brown or even bright yellow, when it comes to specimens from Maio Island, which may actually represent a distinct subspecies. In some specimens, narrow spiral bands in lighter shades of brown can be observed. Rarely with irregular white blotches. Aperture is white. Shell length ranges from 25 to 28 mm. Like all species within the genus Conus, these snails are predatory and venomous. They are capable of "stinging" humans and therefore live ones should be handled carefully or not at all.

Toxin Type: Conopeptides; venom type not reported.

Conus joserochoi (Cossignani, 2014)

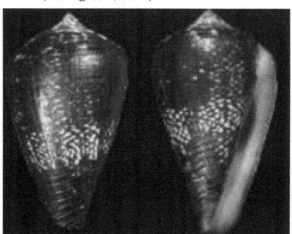

Abapertural view   Apertural view

Common Name: Good shell

Geographical Distribution: Boa Vista – Cape Verde

Habitat: 0.3 to 8 meters deep, among rocks

Identifying Features: Shell of this species is pear-shaped with medium high spire, which is slightly concave or convex. Shoulder is rounded and aperture is quite wide and whose adapical join creates an imperceptible step. Color of the inside of aperture is white–blue and almost homogeneous. Protoconch is low and domed in line with the early whorls of the spire. Coloring of the spire is tawny–brown to dark brown with sparse white spots. Whorl tops are crossed by 4 well-defined spiral furrows. Last whorl has dark reddish–brown coloring in harmony with the color of the spire with a webbing pattern evident that manifests itself in clearly in the central median spiral strip, which occupies a fifth of the last whorl. Shell size ranges from 18 to 35 mm. Like all species within the genus Conus, these snails are predatory and venomous. They are capable of "stinging" humans and therefore live ones should be handled carefully or not at all.

Toxin Type: Conopeptides; venom type not reported.

Conus jourdanida (Motta, 1984)
Image not available

Common Name: Not reported

Geographical Distribution: St Helena

Habitat: Under stones

Identifying Features: Shell of this species is small to moderately small, very slightly pyriform, wide at the shoulder and with a moderately high conical spire, which has a convex profile. Shoulder is subangulated. Sutural ramps are with spiral striae. Ground color of the shell is chestnut brown, decorated just below the shoulder with a necklace of horizontal oblong speckles of a light blue shade, followed, at its periphery, by a narrow band of the same color and is irregularly spotted with brown dots. Another, wider bluish band appears just below mid body and the upper half of it is marked with irregular brown patches. Anterior end of the body whorl also shows rows of bluish speckles. Aperture is bluish-white, whereas the spire is of the same color as the body whorls. Shell size is 28.8 × 18.5 mm. Like all species within the genus Conus, these snails are predatory and venomous. They are capable of "stinging" humans and therefore live ones should be handled carefully or not at all.

Toxin Type: Conopeptides; venom type not reported.

Conus jucundus (Sowerby III, 1887)

Abapertural view   **Apertural view**

Common Name: Abbott's cone

Geographical Distribution: Caribbean

Habitat: Shallow reefs

Identifying Features: Shell of this species is of moderate weight and thick, with good gloss. It is also pyriform or low conical. Shoulder is wide tapering strongly to base. Broad spiral ridges are granulose and are extending posteriorly. Spire is low and its sides are straight and sometimes weakly stepped. Spire whorls are with large, low indistinct coronations. Body whorl is red pink or pale red brown, occasionally olive dark brown. White spiral bands are located at base, midbody and shoulder. White blotches containing brown spots and streaks are seen. Larger brown spots and axial flammules are also occasionally seen. Spire is with alternating blotches of white and red brown suffused pink. Aperture is moderate and uniform. Outer lip is concave and fragil. Mouth fades are violet to pinkish rose. It has strong brown spots and streaks developed as axial flammules extending over large portion of body and not restricted to white spiral bands. Body whorl color is reddish brown or brown; mouth white. Shell size is 35 x 20 mm. Like all species within the genus Conus, these snails are predatory and venomous. They are capable of "stinging" humans and therefore live ones should be handled carefully or not at all.

Toxin Type: Conopeptides; venom type not reported.

Conus juliae **(Clench, 1942) (=** *Conus amphiurgus***)**

Apertural view

Common Name: Julia Clench's cone

Geographical Distribution: North Carolina, Florida; Florida: East Florida, West Florida; Jamaica

Habitat: No Data

Identifying Features: Shell of this species is moderately heavy, with low/ high gloss. It is low biconical and sides are usually convex. Spiral ridges are seen basal. Body whorl is with minute spiral/axial thread. Shoulder is roundly angulate and slightly concave above. Spire is low, moderate and is sharply pointed. Sides are straight/concave. Tops of whorls are flat/ concave. Body whorl has some tone of red pink orange or even yellow. A whitish midbody band, which is rarely olive green or tan is sometimes infiltrated with irregular brownish blotches. Usually, a row of brown square blotches is seen at anterior of band. Spire and shoulder are white with red to yellow blotches. Margin of shoulder is with dark brown blotches. Tip spire is pink. Aperture is moderately wide more anteriorly and outer lip is thin straight/concave. Mouth is white pinkish faint violet. Size of the shell is 54 × 27 mm. Like all species within the genus Conus, these snails are predatory and venomous. They are capable of "stinging" humans and therefore live ones should be handled carefully or not at all.

Toxin Type: Conopeptides; venom type not reported.

Conus julieandreae **(Cargile, 1995)**

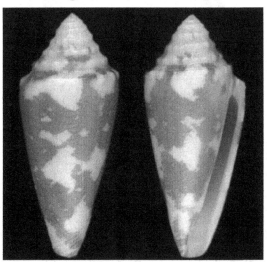

Abapertural view   Apertural view

Common Name: Sea snail

Geographical Distribution: Southern Belize to Santa Marta, Colombia

Habitat: At depths of between 3 and 30 m where it is found in coral sand near reefs.

Identifying Features: This species has a small shell with a spire which is coronated and moderately high. Sides are slightly convex to slightly concave. Shoulder is angulate and has nodules. Body whorl is mildly convex, with a glossy surface. Base color of the shell is white to yellowish white, overlaid with irregular reddish brown to dark brown blotches which form two indistinct spiral bands above and below mid-body, with the upper band often continuous with flammules on the spire whorls. On the anterior half of the body whorl, beads and dots are coincident with the pustulose sculpture. Aperture color is white. Size of the shell is 23 x 10 mm. Like all species within the genus Conus, these snails are predatory and venomous. They are capable of "stinging" humans and therefore live ones should be handled carefully or not at all.

Toxin Type: Conopeptides; venom type not reported.

Conus julii **(Lienard, 1870)**

Apertural view   **Abapertural view**

Common Name: Mascarene cone

Geographical Distribution: Reunion Island (Mauritius)

Habitat: At depths of 25–100 m

Identifying Features: Shell of this species is medium-sized to moderately large and moderately solid-to-solid. Last whorl is ovate to ventricosely conic. Outline is almost straight at adapical fourth, then convex and rather angulate at position of maximum diameter, almost straight below center. Aperture is broad at base, and narrow near shoulder. Shoulder is angulate to subangulate. Spire is low and its outline is slightly concave, straight or sigmoid. Larval shell is of 3–3.5 whorls with a maximum diameter 0.8–0.9 mm. About first 3.5 postnuclear whorls are tuberculate. Teleoconchsutural ramps are slightly concave, with 0 increasing to 1–2 spiral grooves in early whorls, grading to 10 weak spiral grooves or numerous often obsolete spiral striae in latest whorls. Last whorl is with prominent widely spaced spiral ribs on basal fourth to third. Ground color is white, sometimes with sparse pinkish violet shadows. Last whorl is with orange to reddish or dark brown wavy axial lines, concentrated or fusing into blotches and forming spiral bands below shoulder, just above center and within basal third. Larval whorls are white. Postnuclear sutural ramps are with orange, violet, or brown radial streaks and blotches; the latter may contain darker radial lines. Aperture is pink to orange behind a white collabral zone. Shell size ranges from 44 to 62 mm. Like all species within the genus Conus, these snails are predatory and venomous. They are capable of "stinging" humans and therefore live ones should be handled carefully or not at all.

Toxin Type: Conopeptides; venom type not reported.

Conus korni (**G. Raybaudi Massilia, 1993) (=** *Pseudolilliconus korni*)

Apertural view   Abapertural view

Common Name: Sea snail

Geographical Distribution: From Djibouti to South Somalia in the Gulf of Aden to Mogadishu

Habitat: At depths between 50 and 150 m

Identifying Features: Shell of this species is very small and light. Last whorl is usually conical to broadly conical. Outline is convex adapically and straight (right side) or concave (left side) below. Shoulder is sharply angulate to carinate. Spire is usually of moderate height and stepped. Outline is concave. Larval shell is of 2.25–2.5 whorls, with fine radial ridges more prominent toward teleoconch. Teleoconch sutural ramps are concave, often with 1–3 spiral grooves in early whorls. Spiral sculpture is obsolete in late whorls. Last whorl is with spiral ribs on basal third and 1–2 weak spiral grooves just below shoulder. Ground color is white. Last whorl is with a broad, interrupted to solid and orangish to blackish brown spiral band on each side of center, sometimes reduced or absent. Central ground-color band is with an indistinct meshwork of gray background shades, often edged with dark brown to black spots. Closely spaced dashed to solid brown spiral lines are extending from base to shoulder, partially articulated with white dashes. Larval shell is bicolored. Postnuclear sutural ramps are with dark brown radial streaks and blotches crossing outer margins. Aperture is showing exterior color pattern. Shell size ranges from 9 to 13 mm. Like all species within the genus Conus, these snails are predatory and venomous. They are capable of "stinging" humans and therefore live ones should be handled carefully or not at all.

Toxin Type: Conopeptides; venom type not reported.

Conus lamberti (Souverbie, 1877) (= *Darioconus lamberti*)

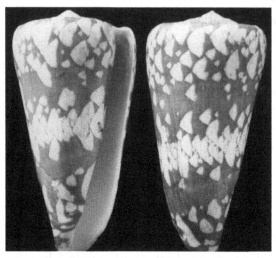

Apertural view　　**Abapertural view**

Common Name: Not designated

Geographical Distribution: New Caledonia region

Habitat: In 30–100 m; on sloping sand bottoms

Identifying Features: Shell of this species is moderately large to large and solid. Last whorl is conical. Outline is convex below shoulder, slightly concave centrally, otherwise straight. Shoulder is subangulate. Spire is low and its outline is concave. First 5–6 postnuclear whorls are tuberculate. Teleoconch sutural ramps are concave, with 2 increasing to 3–5 spiral grooves, turning into striae in latest whorls. Last whorl is with a few spiral ribs at base. Color of the shell is light brown to reddish brown. Last whorl is with small and medium-sized, separate or overlapping white tents and flecks, concentrated at center, below shoulder and at base. Postnuclear sutural ramps are with white radial streaks and blotches. Aperture is white. Shell size ranges from 70 to 114 mm. It feeds on other gastropods. Like all species within the genus Conus, these snails are predatory and venomous. They are capable of "stinging" humans and therefore live ones should be handled carefully or not at all.

Toxin Type: Conopeptides; venom type not reported.

Conus laterculatus **(G. B. Sowerby II, 1870)** (= *Asprella laterculata*)

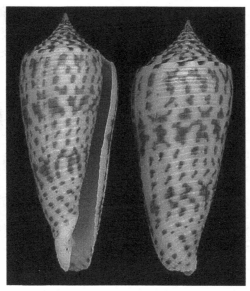

Apertural view   Abapertural view

Common Name: Not reported

Geographical Distribution: Off the Philippines, Borneo and Vietnam

Habitat: At depths between 50 and 240 m.

Identifying Features: Shell of this species is distantly channeled through-out, the interstices usually plane, sometimes minutely granular. Channels are narrow and longitudinally striated. Spire is much elevated, acuminated, striate and sometimes obscurely minutely coronated. Color of the shell is yellowish brown, with light chestnut longitudinal short irregular lines, and clouds of the same color forming three obscure interrupted bands. Size of an adult shell varies between 33 mm and 64 mm. Like all species with the superfamily Conoidea, these snails are predatory and venomous. They are capable of "stinging" humans; therefore, live ones should be handled carefully or not at all.

Toxin Type: Conopeptides; venom type not reported.

Conus lemniscatus **(Reeve, 1849)** **(=** *Lamniconus lemniscatus***)**

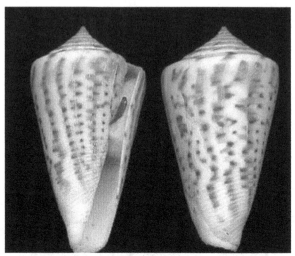

Apertural view   **Abapertural view**

Common Name: Ribbon cone

Geographical Distribution: Caribbean Sea and in the Western Atlantic Ocean along Brazil and Argentina

Habitat: At depths between 35 and 120 m in muddy sand

Identifying Features: Shell of this species shows slightly contracted sides. Spire is acuminated with strong growth lines. Body whorl is delicately ridged throughout. Color of the shell is whitish, maculated with chestnut, and with every alternate ridge chestnut-spotted. Size of an adult shell varies between 20 mm and 65 mm. Like all species within the genus Conus, these snails are predatory and venomous. They are capable of "stinging" humans and therefore live ones should be handled carefully or not at all.

Toxin Type: Conopeptides; venom type not reported.

Conus lenavati (da Motta, 1982) (= *Kioconus lenavati*)

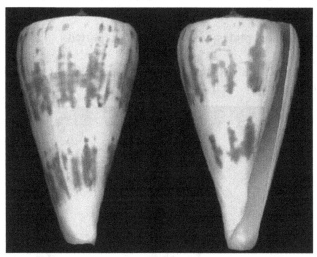

Abapertural view   **Apertural view**

Common Name: Sea snail

Geographical Distribution: Philippines and South China Sea.

Habitat: At depths of between 100 and 240 m

Identifying Features: Shell of this species is moderately large to large and solid. Last whorl is usually conical or ventricosely conical and often pyriform. Outline is convex at adapical third to two-thirds, straight to concave below. Shoulder is angulate. Spire is usually low and outline is concave. Larval shell is of about 3.5 whorls with a maximum diameter 0.9–1.0 mm. First 2.5–4.5 postnuclear whorls are tuberculate. Teleoconch sutural ramps are flat, with 1–2 increasing to 4–6 spiral grooves. Last whorl is with weak spiral ribs and/or ribbons at base. Ground color is white. Last whorl is with brown axial blotches on adapical three-fourth and an incomplete spiral band of the same color on each side of center. Central ground-color band is usually continuous; subcentral pattern elements may be absent. Base is usually white, occasionally variably tinged with cream. Larval whorls are white. Teleoconch sutural ramps are with irregularly set brown axial markings. Aperture is white. Shell size ranges from 55 to 91 mm. Like all species within the genus Conus, these snails are predatory

and venomous. They are capable of "stinging" humans and therefore live
ones should be handled carefully or not at all.

Toxin Type: Conopeptides; venom type not reported.

Conus leobrerai **(da Motta and Martin, 1982)** (= *Phasmoconus leobrerai*)

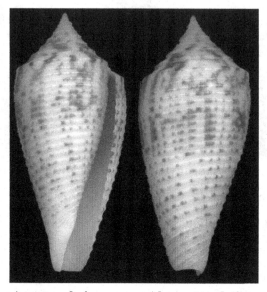

Apertural view   Abapertural view

Common Name: Not known

Geographical Distribution: Philippines, mainly in Cebu, Negros and
Bohol. Probably in Solomon Islands

Habitat: At depths of between 35 and 250 m where they live in deep sand
and/or mud.

Identifying Features: Shell of this species is moderately small, moderately
light to moderately solid. Last whorl is ventricosely conical-to-conical.
Outline is convex adapically, straight (right side) or concave (left side)
below. Shoulder is angulate. Spire is of moderate height to high and is
slightly stepped. Outline is concave. Larval shell is of 2–2.25 whorls with
a maximum diameter 0.8–0.9 mm. First 1–2 postnuclear whorls are often
weakly tuberculate. Teleoconch sutural ramps are flat to slightly concave,

with 0–1 increasing to 4–7 spiral grooves. Last whorl is with spiral ribs from base to shoulder, sometimes ribs basally and ribbons adapically. Grooves between axially striate are containing a fine spiral rib and/or 1–3 spiral threads. Ground color is white. Last whorl is with spiral rows of brown dots on ribs; brown dots often-reduced in large specimens. Dots are fusing into spots and axial streaks below shoulder, above center and occasionally also below center, forming 1–2 or sometimes 3 spiral bands. Larval whorls are white. Postnuclear sutural ramps are with brown radial streaks and blotches. Aperture is white. Shell size ranges from 25 to 35 mm. Like all species within the genus Conus, these snails are predatory and venomous. They are capable of "stinging" humans and therefore live ones should be handled carefully or not at all.

Toxin Type: Conopeptides; venom type not reported.

***Conus leviteni* (Tucker, Tenorio and Chaney, 2011) (= *Darioconus leviteni*)**

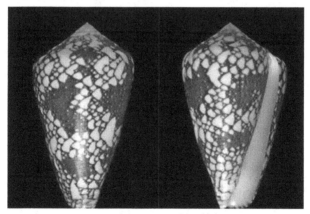

Abapertural view   **Apertural view**

Common Name: Hawaiian penniform cone

Geographical Distribution: Endemic to Hawaii

Habitat: Sand under rocks in tide pools to depths of 5 m

Identifying Features: This species has medium-sized, solid shells which are ventricosely conical with a low to moderate spire. Sutural ramps are flat to concave, with two or three weak cords disappearing in later whorl.

Protoconch is paucispiral, pink, mamillated. Ground color of the shell is white. Last whorl and spire are patterned with different shades of brown overlaid with white tent markings. Aperture is white. Shell size ranges from 35 to 61 mm. Like all species within the genus Conus, these snails are predatory and venomous. They are capable of "stinging" humans and therefore live ones should be handled carefully or not at all.

Toxin Type: Conopeptides; venom type not reported.

Conus lienardi **(Bernardi et Crosse, 1861) (=** *Graphiconus lienardi***)**

Apertural view   Abapertural view

Common Name: Lienard's cone

Geographical Distribution: Indo-Pacific species; New Caledonia, Solomon Island, Vanuatu, and Philippines; probably also Amirante Island.

Habitat: In 3–60 m on sand

Identifying Features: Shell of this species is medium-sized to moderately large and moderately light to moderately solid. Last whorl is usually ventricosely conical or conical. Outline is convex adapically, straight below. Left side is somewhat concave near base. Shoulder is angulate to subangulate. Spire is usually of moderate height and outline variably is concave. Larval shell is of about 2.25–2.50 whorls, with a maximum diameter 0.8–0.9 mm. First 3–7 postnuclear whorls are tuberculate. Teleoconch sutural ramps are flat to slightly concave, with 0–1 increasing to 3–4 spiral grooves. Latest ramps are with faint or obsolete grooves but many additional spiral striae. Last whorl is with widely spaced spiral grooves on basal third or half. Ground color is white. In light-colored specimens, last whorl is with a generally fine and often incomplete, brown or orange reticulate pattern and variously sized flecks aligned in a spiral row on each side of center. Sometimes pattern is only of scattered curved or zigzag-shaped axial dashes. In dark-colored specimens, last whorl is covered with large dark violet-brown zones and bluish background shadows leaving only sparse blotches and tents of white. Larval whorls are light brown. About first 2 postnuclear sutural ramps are brown. Following spire whorls are matching last whorl in color pattern. Light-colored shells are with a white aperture, pale pink or violet; dark-colored shells with a bluish white aperture. Shell size ranges from 35 to 63 mm. This species preys on worms. Like all species within the genus Conus, these snails are predatory and venomous. They are capable of "stinging" humans and therefore live ones should be handled carefully or not at all.

Toxin Type: Conopeptides; venom type not reported.

Conus lindae **(Petuch, 1987) (=** *Conus ignotus, Lindaconus lindae***)**

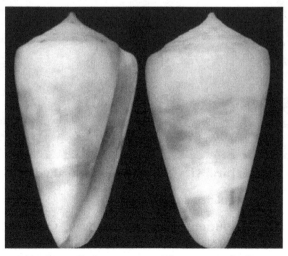

Apertural view   Abapertural view

Common Name: Not known

Geographical Distribution: Endemic to the Bahamas and is found from Grand Bahama Island to Victory Cay, Bahamas, along the western edge of the Bahamas Platform

Habitat: At depths of 400–500 m

Identifying Features: Shell of this species has narrow, widely spaced ridges over the body whorl. Spiral sculpture on the ribbed form of this species is shallow and closely spaced. Protoconch is of 1.25 whorls. Maximum reported size of the shell is 31 mm. Like all species within the genus Conus, these snails are predatory and venomous. They are capable of "stinging" humans and therefore live ones should be handled carefully or not at all.

Toxin Type: Conopeptides; venom type not reported.

Conus lischkeanus **(Weinkauff, 1875) (=** *Calamiconus lischkeanus***)**

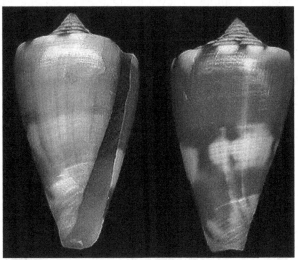

Apertural view   Abapertural view

Common Name: Kermadec cone

Geographical Distribution: Japan and Taiwan, Queensland, Western Australia, New Zealand, New Caledonia, the Kermadec Island, Philippines, Indian Ocean (from Natal, South Africa to the Gulf of Aden).

Habitat: From the intertidal zone to about 200 m; from bare limestone pavement or sand; sometimes among weed or coral rubble

Identifying Features: Whorls of the spire contain a shallow channel. Body whorl is smooth and striate at the base. Color of the shell is sulfur-yellow, without ornamentation except maculations on the spire. Aperture is white. Size of an adult shell varies between 20 mm and 75 mm. Like all species within the genus Conus, these snails are predatory and venomous. They are capable of "stinging" humans and therefore live ones should be handled carefully or not at all.

Toxin Type: Conopeptides; venom type not reported

Conus litoglyphus **(Meuschen, F.C., Hwass, C.H. in Bruguière, J.G., 1792) (=** *Strategoconus litoglyphus***)**

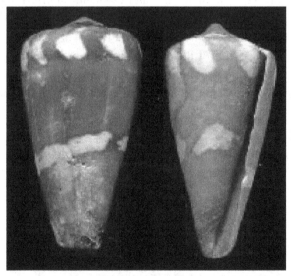

Abapertural view   Apertural view

Common Name: Lithograph cone

Geographical Distribution: Indo-Pacific species (along the Red Sea and in the Indian Ocean along Aldabra, Chagos, the Mascarene Basin and Mauritius).

Habitat: Coral reefs, lagoon pinnacles and exposed rocky platforms; turf algae, macro algae and sand substrates

Identifying Features: Shell of this species is medium- sized to moderately large and moderately solid-to-solid. Last whorl is usually conical. Outline is straight and concave below shoulder. Shoulder is angulate. Spire is low. Outline is usually concave and occasionally convex in early whorls. Larval shell is multispiral with a maximum diameter about 0.8 mm. Early postnuclear whorls are tuberculate. Teleochonce sutural ramps are flat, with 1–3 spiral grooves disappearing in late whorls. Last whorl is with coarse, widely spaced, partially granulose spiral ribs basally and sometimes replaced by obsolete smooth ribs or followed by scattered spiral rows of granules to shoulder. Ground color is white. Last whorl is overlaid with brownish olive or orangish to dark brown, leaving spiral groundcolor bands at shoulder and below center. White bands are solid, regularly or irregularly interrupted by axial color markings, or covered with olive to dark brown. Base is dark brown. Larval whorls and adjacent sutural ramps are gray. Late ramps are with confluent radial blotches matching last whorl pattern in color. Aperture is white or pale brown-dark brown at base. Size of the shell ranges from 40 to 75 mm. Like all species within the genus Conus, these snails are predatory and venomous. They are capable of "stinging" humans and therefore live ones should be handled carefully or not at all.

Toxin Type: Conopeptides; venom type not reported.

Conus locumtenens **(Blumenbach, 1791)** (= *Leptoconus locumtenens*)

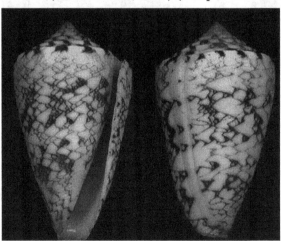

Apertural view   **Abapertural view**

Common Name: Vice admiral cone

Geographical Distribution: Red Sea, Gulf of Aden and in the Indian Ocean along Somalia

Habitat: In 2–10 m; under seaweed, on sand and mud.

Identifying Features: Shell of this species is medium-sized to moderately large and moderately solid-to-solid. Last whorl is usually conical. Outline is either almost straight or convex at adapical third. Shoulder is angulate to slightly carinate. Spire is of low to moderate height and stepped. Outline is concave to straight. Early postnuclear whorls appear weakly tuberculate. Teleoconch sutural ramps are flat to concave towards shoulder. Later ramps are with 2–3 often faint spiral grooves changing into a varying number of weak to obsolete striae. Last whorl is with variably broad, weak spiral ribbons and is often posteriorly edged with a fine spiral rib, on basal third. Ground color is white and is occasionally tinged with blue. Last whorl is with usually fine yellowish to blackish brown reticulations bordering white tents that vary widely in size. Reticulate pattern is often concentrated in 2 or more darker spiral bands of variable width, solidity and color. Within dark bands, reticulated lines may turn into wavy axial lines. Postnuclear sutural ramps are with reddish to blackish brown radial streaks and blotches that cross the outer margins. Aperture is white, and violet-brown deep within. Shell size ranges from 35 to 66 mm. Like all species within the genus Conus, these snails are predatory and venomous. They are capable of "stinging" humans and therefore live ones should be handled carefully or not at all.

Toxin Type: Conopeptides; venom type not reported.

Conus madagascariensis (Sowerby II, 1857)

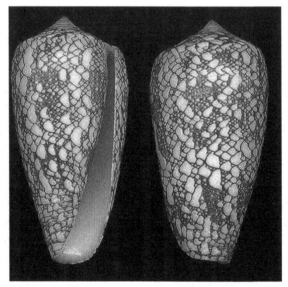

Apertural view   Abapertural view

Common Name: Not known

Geographical Distribution: Endemic to South India along both the western and eastern coasts

Habitat: Infralittoral fringe to depths of about 20–50 m; shallow-subtidal reef flats; coarse sand and rubble; sea-weed and beneath rocks

Identifying Features: Shell of this species is medium-sized to moderately large and moderately solid-to-solid. Last whorl is ventricosely conical to conoid-cylindrical. Outline is moderately convex at adapical third and less so to straight below. Left side is often slightly concave near base. Aperture is somewhat wider at base than near shoulder. Shoulder is angulate. Spire is low and its outline is slightly concave to slightly sigmoid, with a straight-sided apex. Larval shell is of about 2 whorls with a maximum diameter about 0.9 mm. Teleoconch sutural ramps are flat to slightly concave, with 3–4 weak to obsolete spiral grooves and many spiral striae in late whorls. Last whorl is with weak spiral ribs on basal third. Ground color is white and is often variably tinged with violet, sometimes more prominently so at base. Last whorl is with a rather fine and regular network of dark brown lines edging numerous tiny to medium-sized ground color tents. Light brown to reddish brown spiral streaks, spots, flecks or blotches are generally arranged in an interrupted spiral band on each side of center and interspersed with spiral lines of alternating darker brown and white markings. Larval shell is white. Early postnuclear sutural ramps are immaculate white to pink. Following ramps are matching last whorl in color pattern. Aperture is white. It is mollusciverous and feeds on other gastropods. Shell size ranges from 45 to 69 mm. Like all species within the genus Conus, these snails are predatory and venomous. They are capable of "stinging" humans and therefore live ones should be handled carefully or not at all

Toxin Type: Conopeptides; venom type not reported.

Conus magnottei (Petuch, 1987) (= *Conus (Purpuriconus) edwardpauli*)

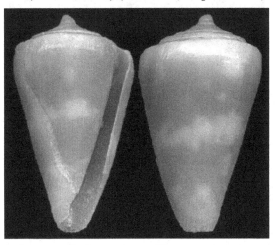

Apertural view   Abapertural view

Common Name: Sea snail

Geographical Distribution: From southern Belize to southern Honduras, and Roatan Island

Habitat: Coral sand and in coral rubble from 1 to 15 m

Identifying Features: Shell of this species is squat, stout, shiny, with high polish. Spire is high, protracted and stepped. Shoulder is sharply angled and carinated. Shoulder carina is undulating and is obsoletely coronated.

Shell color is bright pinkish-salmon with narrow, pale whitish-pink band around midbody. Occasional specimens are deep purple-blue in color. Maximum size of the shell is 23 mm only. Like all species within the genus Conus, these snails are predatory and venomous. They are capable of "stinging" humans and therefore live ones should be handled carefully or not at all.

Toxin Type: Conopeptides; venom type not reported.

Conus maioensis **(Trovao, Rolan and Feliz-Ales, 1990) (=** *Africonus maioensis***)**

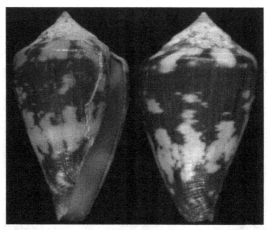

Apertural view   Abapertural view

Common Name: Not reported

Geographical Distribution: Eastern Atlantic Ocean and in the Mediterranean Sea

Habitat: Depths of 1 to 3 m under rocks

Identifying Features: Shell of this species is dark brown with a midbody band of large, irregular, bluish blotches. Shell has a more raised spire, with a preponderance of light colored blotches. Adults of the species typically grow to 36 mm in length. Like all species within the genus Conus, these snails are predatory and venomous. They are capable of "stinging" humans and therefore live ones should be handled carefully or not at all.

Toxin Type: Conopeptides; venom type not reported.

Conus malacanus (Hwass in Bruguière, 1792) (= *Stellaconus malacanus*)

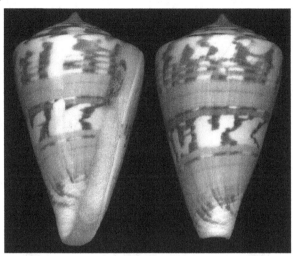

Apertural view   Abapertural view

Common Name: Malacca cone

Geographical Distribution: Off southern India, Sri Lanka, and Papua New Guinea

Habitat: At depths of approx. 5 to 55 m; mainly on sand

Identifying Features: Shell of this species is medium sized to moderately large and solid to heavy. Body whorl is conical to broadly conical. Outline is variably convex at adapical third, straight below. Shoulder is broad and carinate. Spire is of low to moderate height and outline is concave to straight. Early whorls are forming a small projecting cone. Body whorl is smooth except for a few weak spiral ridges above base. Ground color is white. Body whorl is usually with two variably broad continuous or interrupted brown spiral bands, leaving white zones below shoulder at center and at base. Several widely spaced narrow axial stripes may be continuous from shoulder to base, or interrupted at mid-body. Spire and shoulder are whitish, sparsely or heavily marked with brownish blotches and curved streaks, early whorls pale brown. Aperture is moderately wide and uniform in width. Outer lip is thin, and straight. Aperture is white or sometimes cream. Periostracum is yellow, thin, smooth and translucent.

Adults of the species will grow to 83 mm. Like all species within the genus Conus, these snails are predatory and venomous. They are capable of "stinging" humans and therefore live ones should be handled carefully or not at all.

Toxin Type: Conopeptides; venom type not reported.

Conus martensi (**E. A. Smith, 1884**) (= *Kioconus martensi*)

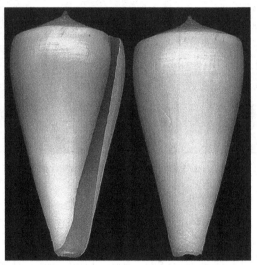

Apertural view　　　　**Abapertural view**

Common Name: Not known

Geographical Distribution: Natal, N.W. Madagascar, Providence Id. (Seychelles), and S. coast of Oman

Habitat: In 40–150 m; zone of sand and sponges along the inner continental shelf

Identifying Features: Shell of this species is medium-sized to moderately large and moderately solid-to-solid. Last whorl is conical and outline is convex at adapical fourth and straight below. Shoulder is angulate. Spire is of low to moderate height and outline is straight to sigmoid or concave. Larval shell is of about 3 whorls with a maximum diameter of about 1 mm. Teleoconch sutural ramps are flat, with 3 increasing to 4–6 spiral grooves. Latest ramps may only have 3–4 grooves. Last whorl is with weak or obsolete spiral ribs at base. Color of the shell is yellow to orange. Last whorl is often with 2 paler spiral bands, at center and at shoulder. Larval whorls are brown. Postnuclear sutural ramps are white, suffused with color tones of last whorl. Aperture is white to pale yellow. Shell size varies from 40 to 78 mm. Like all species within the genus Conus, these snails are predatory and venomous. They are capable of "stinging" humans and therefore live ones should be handled carefully or not at all.

Toxin Type: Conopeptides; venom type not reported.

Conus milneedwardsi (Jousseaume, 1894)

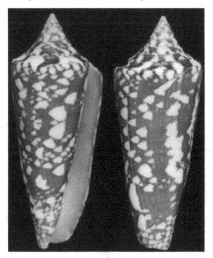

Apertural view   **Abapertural view**

Common Name: Glory of India; spired cone

Geographical Distribution: Along the African coast from KwaZulu-Natal, South Africa, to the Red Sea

Habitat: In 50–180 m; rocky substrate and sand bottoms

Identifying Features: Shell of this species is moderately large to large and moderately solid-to-solid. Last whorl is narrowly conical. Outline is nearly straight. Depth of exhalent notch is about 1/3 to about 2/5 of maximum diameter. Shoulder is angulate to sharply angulate. Spire is stepped and usually high. Outline is generally straight. Maximum diameter of larval shell is about 0.9 mm. First 6–10 postnuclear whorls are tuberculate. Teleoconch sutural ramps are slightly concave-to-concave, often less so in *Conus kawamurai*, with 0–1 increasing to 4–7 spiral grooves; spiral sculpture may be very weak in latest whorls. Last whorl is with variably weak, axially striate spiral grooves near base, separated by ribs at anterior end and by ribbons above. Ground color is white and generally with 2 pink spiral bands on last whorl, just above center and within basal third. Last whorl is generally with reddish brown reticulated lines forming small to large triangular, quadrangular and round markings. Larval shell is white to gray. Teleoconch spire is matching last whorl in color pattern. Aperture is pink to orangish pink deep within. Shell size ranges from 60 to 174 all species within the genus Conus, these snails are predatory and venomous. They are capable of "stinging" humans and therefore live ones should be handled carefully or not at all.

Toxin Type: Conopeptides; venom type not reported.

***Conus miniexcelsus* (Oliviera and Biggs, 2010) (= *Kurodaconus miniexcelsus*)**

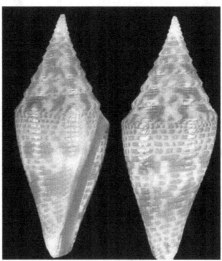

Apertural view   Abapertural view

Common Name: Not reported

Geographical Distribution: Philippines; Japan

Habitat: Not known

Identifying Features: Shell of this species is medium sized, thick and moderately light, glossy. Body whorl is elongate and cylindrical. Shoulder is indistinct. Spire is high. Outline is convex and sharply pointed. Earliest two to three whorls are with fine nodules, and later whorls are with indistinct spiral ridges often bearing small granules. Body whorl is with granulose spiral ribs from base to anterior half of the body whorl. Aperture is narrow and slightly widened anteriorly. Outer lip is thick and slightly convex. Ground color is yellowish. Body whorl is with dark brown. Blurred axial streaks and blotches are seen. Pattern elements are fusing into a spiral band at center and similar but narrower band above base. Apex is white. Shoulder edges of early post nuclear whorls including tubercles are with brown band. Aperture is white. Shell size ranges from 19 to 37 mm. Like all species within the genus Conus, these snails are predatory and venomous. They are capable of "stinging" humans and therefore live ones should be handled carefully or not at all.

Toxin Type: Conopeptides; venom type not reported.

Conus moluccensis **(Küster, 1838)** (= *Fulgiconus moluccensis*)

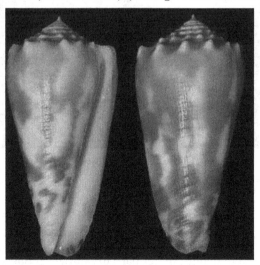

Apertural view **Abapertural view**

Common Name: Molucca cone

Geographical Distribution: Indian Ocean along the Mascarene Basin

Habitat: Sand, gravel, or coral rubble substrate on ledges and in sand pockets at 20 to 240 m

Identifying Features: The coronated shell of this species is yellowish white, marbled and streaked with chestnut, with minute revolving lines of granules, which are often somewhat articulated red-brown and white. Size of an adult shell varies between 30 mm and 60 mm. It has a plankto-trophic development. Like all species within the genus Conus, these snails are predatory and venomous. They are capable of "stinging" humans and therefore live ones should be handled carefully or not at all.

Toxin Type: Conopeptides; venom type not reported.

Conus mordeirae **(Rolán and Trovão in Rolán, 1990)**

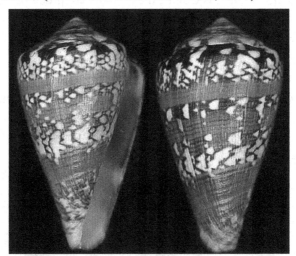

Apertural view   Abapertural view

Common Name: Not reported

Geographical Distribution: Endemic to the Cape Verde Islands where it is found only in the Baía da Mordeira in the south-west of the island of Sal

Habitat: At 1 to 5 m depths; either partially buried in fine sandy or gravel bottom under rocks and dead coral slabs, or partially covered with fine muddy sand among small stones and green filament algae

Identifying Features: Color of the shell is greenish yellow to olive green. It has a white spire (occasionally somewhat elevated) with dark brown blotches which extent to the shoulder, which is also lightly colored. There are often present two bands of white blotches with dark brown streaks distributed above and below the midbody of the last whorl. In some instances, the dark brown color is absent, and only the white blotches are present on a greenish or greenish-brown background. Adults of the species typically grow to 35 mm in length. Like all species within the genus Conus, these snails are predatory and venomous. They are capable of "stinging" humans and therefore live ones should be handled carefully or not at all.

Toxin Type: Conopeptides; venom type not reported.

Conus moreleti Crosse, 1858 (= *Conus (Virgiconus) moreleti*)

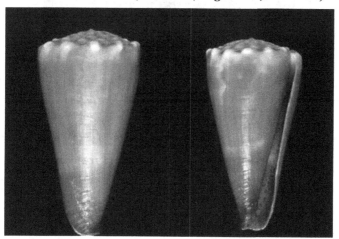

Abapertural view   Apertural view

Common Name: Morelet's cone.

Geographical Distribution: Across the central Indian Ocean, in the West Pacific and off the Hawaiian Islands

Habitat: Coral reefs between 1 and 50 m deep; fringing reefs, which experience strong tidal flow

Identifying Features: Shell of this species is smooth, slender and olive green without distinct mid-body spiral or granules. Spire is low and bumpy. Shell size ranges from 35 to 45 mm. Animal reddish brown with distinctive orange, yellow, and black markings. Like all species within the genus Conus, these snails are predatory and venomous. They are capable of "stinging" humans and therefore live ones should be handled carefully or not at all.

Toxin Type: Conopeptides; venom type not reported.

Conus mozambicus (Hwass in Bruguière, 1792)

Apertural view   Abapertural view

Common Name: Elongate cone

Geographical Distribution: Off the southern African coast from Lüderitz Bay to Mossel Bay

Habitat: Subtidally in shallow water (0–60 m); crevices in low tide reef platforms, usually half-buried in sand in low tide pools

Identifying Features: This species has a medium-sized shell, which may grow to 65 mm in total length. It has a sharply pointed spire. Shell color is dull and mottled with brown, and there may be darker blotches at the shoulder. Spire of the shell is stepped. It feeds on polychaete worms. Like all species within the genus Conus, these snails are predatory and venomous. They are capable of "stinging" humans and therefore live ones should be handled carefully or not at all.

Toxin Type: Conopeptides; venom type not reported.

Conus mucronatus **(Reeve, 1843)** *(= Phasmoconus mucronatus)*

Apertural view   Abapertural view

Common Name: Deep-grooved cone

Geographical Distribution: Indian Ocean along the Mascarene Basin; in the Pacific Ocean along the Philippines to Papua New Guinea, Solomon Islands, Queensland, Australia, and Vanuatu; along India and in the South China Sea

Habitat: Usually in 3–50 m, more common below 20 m

Identifying Features: Shell of this species is moderately small to medium-sized and moderately light to moderately solid. Last whorl is conical to ventricosely conical. Outline is convex adapically almost straight below. Left side is slightly concave at base. Basal part of columella sometimes is deflected to left. Shoulder is sharply angulate. Spire is of moderate height and is slightly stepped. Outline is concave. Larval shell is of about 3 whorls with a maximum diameter about 0.7 mm. First 2–4 postnuclear whorls are tuberculate. Teleoconch sutural ramps are almost flat, with 0–1 increasing to 5–9 spiral grooves. Last whorl is with variably spaced, axially striate to punctate spiral grooves below center, occasionally to shoulder, separated by ribs at anterior end and by ribbons above; grooves may contain spiral threads and ribbons may be partially subdivided in 2–3 narrower elevations. Ground color is white. Last whorl is with evenly spaced yellowish to reddish brown dotted, dashed or solid spiral lines. Axial streaks or blotches occasionally form a spiral band above center and traces of bands at shoulder and near base. Larval whorls are white. Shell size ranges from 33 to 50 mm. Like all species within the genus Conus, these snails are predatory and venomous. They are capable of "stinging" humans and therefore live ones should be handled carefully or not at all.

Toxin Type: Conopeptides; venom type not reported.

Conus muriculatus **(G. B. Sowerby II, 1833) (=** *Lividoconus muriculatus***)**

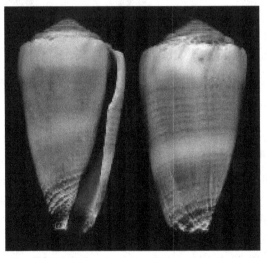

Apertural view   Abapertural view

Common Name: Muricate cone

Geographical Distribution: Indian Ocean along the Mascarene Basin to Western Australia; in the Pacific Ocean from Japan to New Caledonia, Fiji, and French Polynesia

Habitat: Intertidal to about 150 m

Identifying Features: The solid shell of this species has straight sides, and a short conical spire. Shoulder is sharply angulated and tuberculated. Body whorl is strongly striate towards the base, encircled throughout with lines of granules. Color of the shell is white, violet-tinged towards the base, with two light chestnut or yellowish brown, broad, irregular and somewhat indistinct bands. Size of an adult shell varies between 15 mm and 50 mm. Like all species within the genus Conus, these snails are predatory and venomous. They are capable of "stinging" humans and therefore live ones should be handled carefully or not at all.

Toxin Type: Conopeptides; venom type not reported.

Conus nanus **(G.B. Sowerby I & G.B. Sowerby II, 1833)**

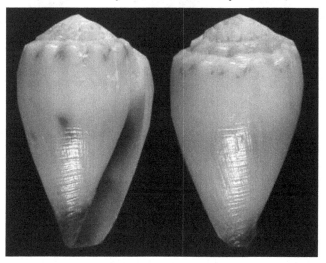

Apertural view　　**Abapertural view**

Common Name: Dwarf cone

Geographical Distribution: Indian Ocean along the Mascarene Basin; in the Indo-Pacific Region (Polynesia, Australia) South Africa

Habitat: Uncommon on intertidal benches, tide pools, and reefs

Identifying Features: Shell of this species is coronated, with a rather depressed spire, granular striae towards the base. Ground color of last whorl is slightly bluish-white with color pattern reduced to a few flecks, a few dotted or dashed spiral lines, or completely absent. Spire color pattern is reduced to dots between tubercles or absent. Aperture is light violet, brown or blue. Size of an adult shell varies between 12 mm and 34 mm. Like all species within the genus Conus, these snails are predatory and venomous. They are capable of "stinging" humans and therefore live ones should be handled carefully or not at all.

Toxin Type: Conopeptides; venom type not reported.

Conus neptunus **(Reeve, 1843)**

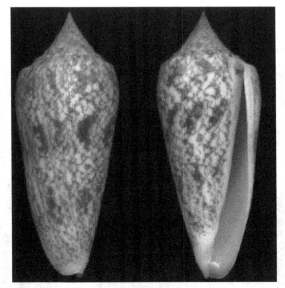

Abapertural view   Apertural view

Common Name: Neptune cone

Geographical Distribution: Along the Philippines, Australia and in the South-west Pacific Ocean

Habitat: Depths of 50–240 m

Identifying Features: Shell of this species is medium-sized and moderately solid. Last whorl is ovate. Outline is convex, and slightly constricted at the base. Shoulder is angulate. Spire is of moderate height and its outline is deeply concave. Protoconch is with more than two whorls and a maximum diameter 0.8 mm. Teleoconch is with 10.5 whorls, which are slightly stepped. Sutural ramp is concave. First 8 whorls are tuberculate. Upper part of last whorl is smooth and basal part is with 8–10 strong spiral ribs. Ground color is cream with irregularly scattered dark brown, curved or angular axial dashes and triangular spots, concentrated and underlain light brown or light violet flecks on both sides of center. Protoconch and base are white. Aperture is light violet. Size of an adult shell varies between 43 mm and 80 mm. Like all species within the genus Conus, these snails are predatory and venomous. They are capable of "stinging" humans and therefore live ones should be handled carefully or not at all.

Toxin Type: Conopeptides; venom type not reported.

Conus nimbosus **(Hwass in Bruguière, 1792) (=** *Rolaniconus nimbosus***)**

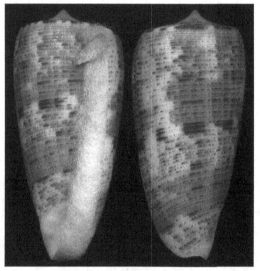

Apertural view   Abapertural view

Common Name: Stormy cone

Geographical Distribution: Indian Ocean along Mozambique, Madagascar, the Seychelles, India, and Sri Lanka; in the Pacific Ocean along Papua New Guinea, Vanuatu, and Samoa

Habitat: Infralittoral reefs to 60 m, on sand flats

Identifying Features: Shell of this species is large, subpyriforrn, broad across shoulder, heavy and thickened. Shell profile is with straight sides. Spire is low and flattened, with only early whorls projecting above plane of spire. Shell is widest below (anterior of) shoulder. Body whorl is shiny, with silky texture and ornamented with extremely numerous, closely packed fine spiral threads. Anterior one-third of shell is sculptured with

20–30 large, strong spiral cords. Base color is varying from lemon-yellow to golden-yellow and is overlaid with 2–3 wide bands of darker yellow or golden-yellow. Some individuals are marked with fine, dark golden-tan spiral lines, often arranged in two bands. Anterior tip is darker yellow or golden-tan. Spire whorls are flat and planar, with early whorls, which are slightly elevated and subpyramidal. Whorls are ornamented with 6 large spiral cords and finer spiral threads between each pair of large cords. Color is varying from pale yellow to golden-yellow. Early whorls are pale brown. Shoulder is sharply angled and edged by large, rounded carina. Sub-sutural area is flattened. Aperture is uniformly wide and slightly flaring toward anterior end. Interior is pure white. Size of an adult shell varies between 33 mm and 65 mm. Like all species within the genus Conus, these snails are predatory and venomous. They are capable of "stinging" humans and therefore live ones should be handled carefully or not at all.

Toxin Type: Conopeptides; venom type not reported.

Conus nocturnus (Hwass, C.H. in Bruguière, J.G., 1792)

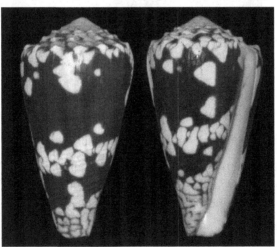

Abapertural view   **Apertural view**

Common Name: Not reported

Geographical Distribution: Moluccas and N.W. New Guinea

Habitat: In 1.5–2 m, on coral sand and on dead coral

Identifying Features: Shell of this species is medium-sized to moderately large and moderately solid-to-solid. Last whorl is slightly ventricosely conical to conoid-cylindrical and outline is convex. Left side is concave abapically. Shoulder is angulate and strongly tuberculate. Spire is of low to moderate height, and outline is straight to slightly concave. Postnuclear whorls are tuberculate. Teleoconch sutural ramps are concave in later whorls, with 2–3 distinct to obsolete spiral grooves. Last whorl is with weak spiral ribs on basal fourth and ribs are often granulose and extending up to shoulder. Ground color is white. Last whorl is with 2 broad blackish brown spiral bands, above and below center. White zones are with dark brown reticulate lines. Teleoconch sutural ramps are with irregularly spaced brown reticulate lines. Aperture is white. Shell size varies from 45 to 86 mm. Like all species within the genus Conus, these snails are predatory and venomous. They are capable of "stinging" humans and therefore live ones should be handled carefully or not at all.

Toxin Type: Conopeptides; venom type not reported.

Conus nussatella **(Linnaeus, 1758) (=** *Hermes nussatellus***)**

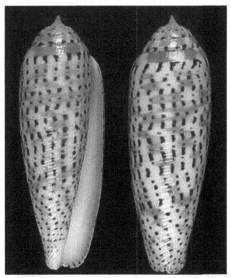

Apertural view   Abapertural view

Common Name: Nussatella cone

Geographical Distribution: Hawaii and Indo-Pacific

Habitat: Sublittoral zone; rare in silt under slabs or rubble at scuba depths

Identifying Features: Shell of this species is medium-sized and moderately solid. Body whorl is narrowly cylindrical. Outline is straight and parallel-sided at upper two thirds. Shoulder is indistinct. Spire is of moderate height and outline is convex. Body whorl is with fine, weakly granulose spiral ribs from base to shoulder. Intervening grooves are spirally striate. Aperture is wide at base than near shoulder. Outer lip is straight, sharp and thick. Ground color is light cream with spiral rows of small dark brown spots and variably sized orange brown axial blotches coalescing axially as well as spirally, especially concentrating above and below the center. Aperture is white. Adult shell varies between 35 mm and 95 mm. It feeds upon other snails. Like all species within the genus Conus, these snails are predatory and venomous. They are capable of "stinging" humans and therefore live ones should be handled carefully or not at all.

Toxin Type: Conopeptides; venom type not reported.

Gradiconus nybakkeni **(Tenorio, Tucker and Chaney, 2012)**

Abapertural view   **Apertural view**

Common Name: Not designated

Geographical Distribution: Bahia de Los Angeles and Bahia los Fragiles, Baja California Sur, Mexico

Habitat: Deep water between 47 and 60 m

Identifying Features: This species has a narrowly conical shell with angular shoulders, an elevated and slightly scalariform spire, and a body whorl with flat sides. Shell shape is not known to vary in this species. Protoconch is paucispiral with two whorls, and fluted nodules are present on the shoulders of the earliest teloconch whorls. Sutural ramps are flat and lack cords. Color consists of light golden brown with white blotches in a spiral pattern. Aperture is blue-white to lavender in color. Protoconch and early whorls are white. Its length is between 12.8 and 50.9 mm. Like all species within the genus Conus, these snails are predatory and venomous. They are capable of "stinging" humans and therefore live ones should be handled carefully or not at all.

Toxin Type: Conopeptides; venom type not reported.

Conus ochroleucus **(Gmelin, 1791)** *(= Graphiconus ochroleucus)*

Apertural view   **Abapertural view**

Common Name: Prefect cone

Geographical Distribution: Pacific Ocean along Taiwan, the Philippines, Papua New Guinea, Indonesia and Fiji; in the Indian Ocean along India

Habitat: Muddy and rocky substrate

Identifying Features: Shell of this species is long and narrow and is distantly grooved towards the base. Its color is yellowish brown and variously shaded, with a rather indistinct median lighter band. White aperture is somewhat wider anteriorly. The striate and acuminate spire is maculated with yellowish brown and white. Size of an adult shell varies between 40 mm and 88 mm. Like all species within the genus Conus, these snails are predatory and venomous. They are capable of "stinging" humans and therefore live ones should be handled carefully or not at all.

Toxin Type: Conopeptides; venom type not reported.

Conus pacificus **(Moolenbeek and Röckel, 1996)** *(= Cylinder pacificus)*

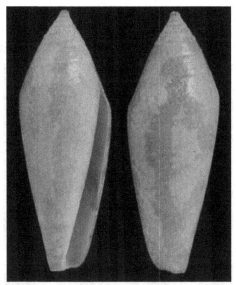

Apertural view   **Abapertural view**

Common Name: Not reported

Geographical Distribution: Only from Futuna Island (Wallis and Futuna)

Habitat: Deep-dwelling species found at 295–600 m., on hard bottom substrate such as gravel and rubble.

There are currently no data in the literature concerning habitats and ecology of this species

Identifying Features: Shell of this species is small, cylindrical, fusiform, smooth, and glossy. Protoconch diameter is 0.8 mm. Teleoconch is with 7.25 whorls, the first three to four whorls with small nodules and two spiral grooves. On the last whorl, there is only one spiral groove. Shoulder is slightly angulate and spire is a little convex. Sutural ramp is almost flat. A few indistinct basal grooves are seen. Ground color is white, with light brown axial streaks and very fine spiral lines consisting of very fine white spots. On the last whorl there are four continuous axial brown streaks from base to suture. Dimensions: shell length 20.2 mm, diameter 7.3 mm, aperture height 14.7 mm. Like all species within the genus Conus, these snails are predatory and venomous. They are capable of "stinging" humans and therefore live ones should be handled carefully or not at all (Moolenbeek and Rockel, 1996).

Toxin Type: Conopeptides; venom type not reported.

Harmoniconus paukstisi **(Tucker, Tenorio and Chaney, 2011)** (= *Conus (Harmoniconus) paukstisi*)

Apertural view   Abapertural view

Common Name: Hawaiian dwarf cone

Geographical Distribution: Throughout the Hawaiian Islands

Habitat: Intertidal benches, tide pools, and reefs

Identifying Features: Shell of this species is small reaching about 35 mm in length. Shell is conical in shape but with slightly convex sides in small shells (< 20 mm). Sides become very convex in larger shells and the shoulders of these are swollen looking with indistinct shoulders. 13 to 22 nodules are present on shells up to 18 mm. Whorl tops of the first 2 or 3 whorls have 2 cords but these increase in number to five or more. Cords spread from the whorl tops over the nodules on to the body whorl. Cords are well developed between adjacent nodules. Cords may be obsolete in the largest specimens. Anterior end and interior of the aperture are colored brown, rarely purple brown. A strong constriction is present inside the aperture at about midbody. Coloration is highly variable depending on growth stage. Smaller shells (< 18 mm) can be all blue-white, marked with brown irregular blotches, have brown spiral lines on the body whorl, or be mottled brown over the body whorl. Spire coloration can be uniform blue-white or have brown markings. Like all species within the genus Conus, these snails are predatory and venomous. They are capable of "stinging" humans and therefore live ones should be handled carefully or not at all.

Toxin Type: Conopeptides; venom type not reported.

Conus patricius **(Hinds, 1843)** (= *Pyruconus patricius*)

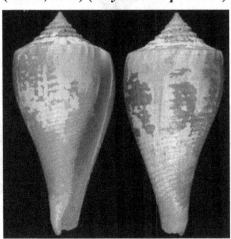

Apertural view   **Abapertural view**

Common Name: Patrician cone, pear shaped cone

Geographical Distribution: Pacific Ocean; from the Gulf of California, Western Mexico, to Ecuador and along the Galápagos Islands

Habitat: At depths between 5 and 50 m on muddy sand

Identifying Features: Size of an adult shell varies between 30 mm and 150 mm. Shell of this species has a light flesh-color. Spire is gently acuminate. Earlier whorls are tuberculated. Body whorl is pyriform. Outline is concave below, with revolving striae towards the base. Like all species within the genus Conus, these snails are predatory and venomous. They are capable of "stinging" humans and therefore live ones should be handled carefully or not at all.

Toxin Type: Conopeptides; venom type not reported.

Conus pertusus **Hwass in Bruguière, 1792 (=** *Conus (Rhizoconus) pertusus)*

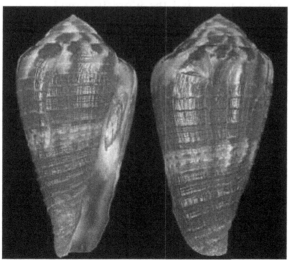

Apertural view   Abapertural view

Common Name: Pertusus cone

Geographical Distribution: Indian Ocean (along Chagos, the Mascarene Basin and Mauritiu); entire Indo-Pacific

Habitat: At depths of 5–120 m; Primarily subtidally under rocks or in rubble by day, emerging on ledges and in caves at night to hunt; intertidal reefs, in sand beneath boulders or subtidal lagoons and interisland reefs and a range of habitat types

Identifying Features: Shell of this species is moderately small and moderately heavy with high gloss. Body whorl is conical. Outline is convex just below shoulder and straight below. Shoulder is angulate. Spire is of moderate height. Outline is convex and apex is sharp. Aperture is narrow, and slightly wider anteriorly. Outer lip is thin, sharp, and straight. Body whorl is smooth except for a few weak spiral ribs at base. Ground color is white and body whorl is with orange red blotches fusing into two variably broad spiral bands on each side of the center. Adapical and central white bands are crossed by axial blotches. Aperture is pale pink. Size of an adult shell varies between 20 mm and 69 mm. This species feeds on polychaetes. Like all species within the genus Conus, these snails are predatory and venomous. They are capable of "stinging" humans and therefore live ones should be handled carefully or not at all.

Toxin Type: Conopeptides; venom type not reported.

Conus pineaui **(Pin and Tack, 1995)**

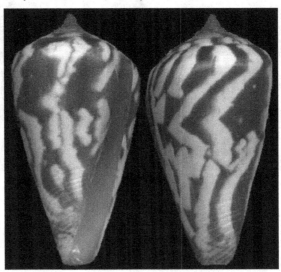

Apertural view   **Abapertural view**

Common Name: Not reported

Geographical Distribution: Endemic to Senegal; from south of the Cap Vert peninsula, Dakar, Goree Island and further south at Petite Côte

Habitat: Deeper water; sandy habitats among rocks; collected through dredging and trawling below 20–30 m depth

Identifying Features: Shell profile of this species is straight. Shell has a narrower shoulder, more elevated spire and rather pyriform shape. Aperture is always white. Adults of the species will grow to 35 mm. Like all species within the genus Conus, these snails are predatory and venomous. They are capable of "stinging" humans and therefore live ones should be handled carefully or not at all.

Toxin Type: Conopeptides; venom type not reported.

Conus plinthis **(Richard and Moolenbeek, 1988) (=*Kioconus plinthis*)**

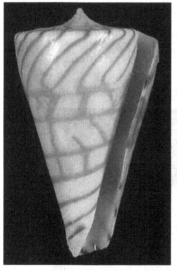 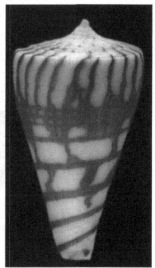

Apertural view   **Abapertural view**

Common Name: Not known

Geographical Distribution: New Caledonia, Norfolk Is., Kermadecs Is. and Northern Three Kings Rise in the Western Pacific

Habitat: At 135–844 m; continental shelves; from sea mounts to knolls

Identifying Features: Shell of this species is moderately small to medium sized and light to moderately solid. Last whorl is conical. Outline is straight and slightly convex at adapical fourth. Shoulder is angulate. Spire is of low to moderate height and its outline is concave. Larval shell is of about 3.25 whorls, with a maximum diameter of 1.1 mm. First 3–4 postnuclear whorls are tuberculate. Teleoconch sutural ramps are flat to slightly concave, with 0–4 increasing to 5–7 spiral grooves. Last whorl is with a few weak spiral ribs and grooves at base. Ground color is white. Last whorl is with about 9 broad brown spiral lines from base to adapical third or fourth. They are widely spaced basally and rather closely spaced adapically. Posterior lines are within a variably broad light to reddish brown spiral band above center. Widely but unevenly spaced brown axial lines and streaks are seen connecting the brown spiral lines and extending from the adapical band to the shoulder ramp. Larval whorls and first 2–3 postnuclear sutural ramps are white to beige. Following sutural ramps are with brown radial lines. Aperture is white. Shell size varies form 20 to 61 all species within the genus Conus, these snails are predatory and venomous. They are capable of "stinging" humans and therefore live ones should be handled carefully or not at all.

Toxin Type: Conopeptides; venom type not reported.

Conus praecellens (A. Adams, 1855) (= *Kurodaconus praecellens*)

Apertural view   **Abapertural view**

Common Name: Admirable cone, Sozon's cone, Chinese cone

Geographical Distribution: In the Indian Ocean along Madagascar, Réunion, Somalia, India, West Thailand and Western Australia; in the Pacific Ocean from Japan to the Philippines and Melanesia (Papua New Guinea, Solomon Islands, New Caledonia, Vanuatu)

Habitat: Muddy ocean beds of depths between 10 and 250 m

Identifying Features: This species has a moderately small shell, which is high-spired. Both spire and body whorl are having a straight outline, making the shell narrowly biconical. Larval shell has 3.0–3.5 whorls which are translucent brownish or purplish. There are 9–11 teleoconch whorls, the first three being ivory-white, without spots, providing a notable contrast to the translucent-colored protoconch. At around the fourth teleoconch whorl, broad brownish spots appear. Ground color is white, with chestnut-brown spots. On the body whorl, there are a series of flat spiral ribbons. Shell pattern on the body whorl can be divided into 3–5 zones. The most posterior, next to the suture, are a series of about 6 spiral ribbons with extremely fine chestnut brown spots. These are followed by a zone with 3 noticeably broader spiral ribbons that have deeper brown and larger spots. The remainder of the shell towards the tip is covered by spiral ribbons that are darker in color and more heavily spotted. Typically, the first 3 to 4 are darker than those towards the anterior end of the shell. Adult shell size range is 20–63 mm. Like all species within the genus Conus, these snails are predatory and venomous. They are capable of "stinging" humans and therefore live ones should be handled carefully or not at all.

Toxin Type: Conopeptides; venom type not reported

Conus pretiosus **(Nevill and Nevill, 1874) (=** *Conus phuketensis***)**

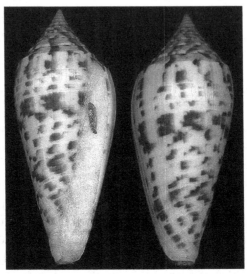

Apertural view   Abapertural view

Common Name: Not found

Geographical Distribution: Off the west coast of Thailand south to Sumatra

Habitat: At depths of between 40 and 120 m

Identifying Features: Shell of this species is moderately large to large and moderately solid. Body whorl is ventricosely conical. Outline is convex adapically and straight below. Shoulder is angulate. Spire is of moderate height and stepped. Outline is straight and apex is pointed and sharp. Body whorl is smooth. Outer lip is thin and sharp. Periostracum is thin and translucent. Ground color is white to cream. Body whorl is with narrow brown (cream to yellowish brown) spiral bands from base to shoulder. Overlying spiral rows are of variously sized and shaped brown to dark brown markings fuse into variably prominent interrupted spiral bands, below shoulder and above center. Spire whorls are white to brown and early whorls are white to pale brown. Postnuclear sutural ramps are white or cream with brown radial streaks and blotches. Aperture is pale pinkish and exterior pattern is visible on the outer lip margin. Periostracum is brown and translucent. Adults of the species will grow to approx. 100 mm.

Like all species within the genus Conus, these snails are predatory and venomous. They are capable of "stinging" humans and therefore live ones should be handled carefully or not at all.

Toxin Type: Conopeptides; venom type not reported.

Conus profundorum **(Kuroda, 1956)** (= *Chelyconus (Profundiconus) profundorum, Conus smirna*)

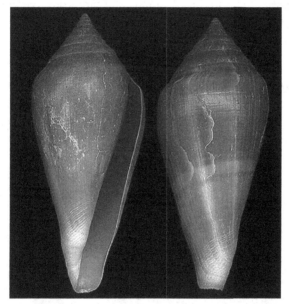

Apertural view   Abapertural view

Common Name: Not reported

Geographical Distribution: Near Kauai, Hawaii; Pacific (Midway Is., New Caledonia and Kermadec Ridge New Zealand); Philippines and Japan.

Habitat: Depth range, 470–570 m.

Identifying Features: Shell of this species is moderately large and moderately solid. Last whorl is slightly pyriform. Outline is convex adapically and slightly concave below. Shoulder is indistinct from spire but with a prominent edge. Spire is high and its outline is almost straight. Larval shell is of more than 2 whorls with a maximum diameter about 1.1 mm.

First 5–6 postnuclear whorls are tuberculate. Teleoconch sutural ramps are flat, with 1–2 increasing to 4–6 spiral grooves, which are obsolete in late whorls. Last whorl is with spiral ribs on basal third, followed by spiral threads to shoulder. Ground color is white. Larval whorls are light brown. Last whorl is with a broad pale brown spiral band on each side of center. Aperture is white. Shell's maximum reported length is 61 mm. Like all species within the genus Conus, these snails are predatory and venomous. They are capable of "stinging" humans and therefore live ones should be handled carefully or not at all.

Toxin Type: Conopeptides; venom type not reported.

***Conus proximus* (G. B. Sowerby II, 1859) (= *Phasmoconus proximus*)**

Abapertural view   Apertural view

Common Name: Sea snail

Geographical Distribution: Indo-Pacific Region (Philippines to Vanuatu and Fiji)

Habitat: In 25–240 m; intertidal; coarse sand at the foot of a reef; coarse sand off reefs

Identifying Features: Shell of this species is moderately small to medium-sized and conical to conoid- cylindrical. Outline is slightly convex. Columella is slightly deflected to left at siphonal fasciole. Shoulder is angulate to subangulate, with about 12–17 tubercles. Spire is of low to moderate height and its outline is concave. Larval shell is of 2 whorls with a maximum diameter 0.8–0.9 mm. Postnuclear whorls are tuberculate. Teleoconch sutural ramps are flat with 0 increasing to 3–4 spiral grooves, sometimes with a few additional striae. Last whorl is with widely to closely spaced, often granulose spiral ribs or ribbons either extending from base to shoulder or restricted to basal part; grooves between punctate or axially striate. Ground color is white to light brown. Last whorl is heavily clouded with yellowish to orangish brown. Color marking is irregularly arranged or forming indistinct spiral bands on both sides of center. Spiral rows are of brown or orange dots and dashes extending from base to shoulder, varying from numerous to absent. Postnuclear sutural ramps are with radial streaks matching last whorl pattern in color, often as pronounced spots between tubercles. Aperture is white to pale blue or violet. Shell size ranges from 25 to 45 mm. Like all species within the genus Conus, these snails are predatory and venomous. They are capable of "stinging" humans and therefore live ones should be handled carefully or not at all.

Toxin Type: Conopeptides; venom type not reported.

Conus pseudocuneolus (**Röckel, Rolán, and Monteiro, 1980**) (= *Africonus pseudocuneolus*)

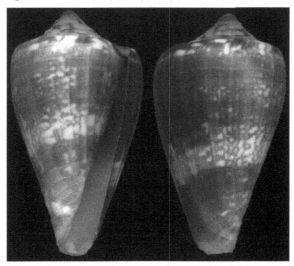

Apertural view   Abapertural view

Common Name: Not reported

Geographical Distribution: In Atlantic Ocean along the Cape Verde islands

Habitat: In sand at shallow depths from 0.5 to 5 m; Adult specimens have been found partially buried in sandy gravel bottom with pink incrusting algae near rock boulders, whereas juveniles prefer shallower water and are normally found on rock platform under rocks.

Identifying Features: Shell profile of this species is quite triangular, with straight sides of the last whorl. Spire is moderately low and conspicuous in some cases. Color is variable, from dark brown to cream brown, passing through all possible intermediate shades. Shell can be patternless, but it most often shows one central band with a netted pattern, and another band lighter than the background color just below the shoulder. Size of an adult shell varies between 13 mm and 40 mm. Like all species within the genus Conus, these snails are predatory and venomous. They are capable of "stinging" humans and therefore live ones should be handled carefully or not at all.

Toxin Type: Conopeptides; venom type not reported.

Conus pulcher **(Lightfoot, 1786)** (= *Kalloconus pulcher*)

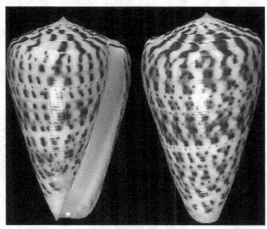

Apertural view   **Abapertural view**

Common Name: Pulcheria cone

Geographical Distribution: Eastern Atlantic (Guinea, Senegal, Angola); Indian Ocean (off the east coast of Africa)

Habitat: Shallow water; sand and mud in a range from 1 m to 50 m

Identifying Features: Shells of this species are conical shaped with a strongly enhanced the upper part of the last turn. Curl is slightly elevated above the last whorl. Siphon outgrowth is short and straight. Mouth is narrow and slightly expanded at the bottom. Outer lip is not extended, strong and smoothly bent in the middle. Sculpture is represented only by growth lines. It has light-shell color, uneven, with some stripes and spots. In the middle of the last whorl takes two broad bands of light, almost white. There are also vertical, narrow light stripes. On the upper side brown spots of more intense shade are seen. Adults are typically growing to 250 mm in length. This species feeds on other mollusks, carrion and sedentary fish. Like all species within the genus Conus, these snails are predatory and venomous. They are capable of "stinging" humans and therefore live ones should be handled carefully or not at all.

Toxin Type: Conopeptides; venom type not reported.

Conus recluzianus **(Bernardi, 1853)**

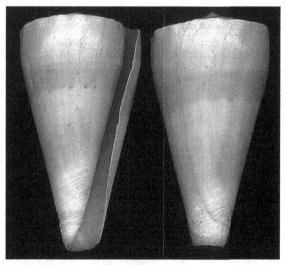

Apertural view   Abapertural view

Common Name: Recluse cone

Geographical Distribution: From S. E. India and Sri Lanka to Singapore, Japan to the Philippines, the Solomon Islands, New Caledonia, Papua New Guinea, Western Australia, Arafura Sea and Queensland

Habitat: At depths between 25–240 m; sandy habitats. Adults of this species are typically between 45–80 mm in length

Identifying Features: Shell of this species is conical in shape, large and porcellaneous in texture. Last whorl is straight. Aperture is narrow, becoming wider below the mid-whorl. Shoulder is angulate in shape and heavily undulate. Spire is low to very low and sutural ramps are straight to slightly concave. They are covered with about 4 strong spiral striae with spiral threads between in and very fine radial growth lines and threads. Body whorl has a base color, which is white with a purple shine. The pattern consists of large axial blotches with in some specimen axial streaks. The irregular yellow to dark brown axial blotches are set into two broad spiral rows. Spire is with numerous radial markings of the same color as the blotches on the body whorl. Basal part of the columella is white, sometimes with a tinge of cream. Shell length varies from 45 to 92 mm. These snails are predatory and venomous. They are capable of "stinging" humans and therefore live ones should be handled carefully or not at all.

Toxin Type: Conopeptides; venom type not reported.

Conus regonae **(Rolán and Trovao in Rolán, 1990)**

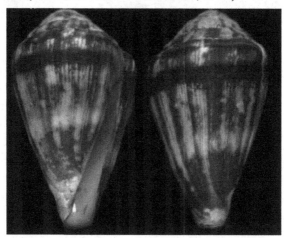

Apertural view   Abapertural view

Common Name: Not known

Geographical Distribution: Endemic to the island of Sal in the Cape Verde Islands.

Habitat: At depths between 2 and 5 m; Very small specimens of this species crawling on the walls of large pools formed during low tide

Identifying Features: Color of the shell of this species is cream brown overlaid with bands of a very dark brown color at the shoulder and at the base, with many axial lines and streaks of the same color distributed over the last whorl. Adults of the species typically grow to 20 mm (range 15–25 mm) in length. These snails are predatory and venomous. They are capable of "stinging" humans and therefore live ones should be handled carefully or not at all.

Toxin Type: Conopeptides; venom type not reported.

Conus retifer **(Menke, 1829)** (= *Conus (Cylinder) rotifer*)

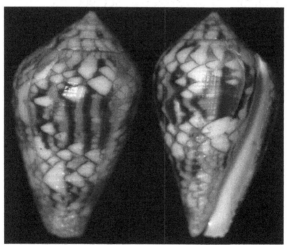

Abapertural view   Apertural view

Common Name: Not known

Geographical Distribution: Mozambique to Tanzania, to Hawaii and French Polynesia.

Habitat: On reef at a depth of 2–3 m

Identifying Features: Shells of this species are moderately small to moderately large and moderately solid-to-solid. Last whorl is ventricosely conical

to ovate or slightly pyriform. Outline is strongly convex adapically, straight
to concave below and more concave on left side. Shoulder is rounded to
almost indistinct. Spire is of low to moderate height and outline is straight
to convex or slightly sigmoid. Larval shell is of about 3.25 whorls with
a maximum diameter of 0.8 mm. First 5–6 postnuclear whorls are tuberculate. Teleoconch sutural ramps are flat to slightly convex, with 1 increasing
to 4–7 spiral grooves. Additional spiral striae are seen on latest ramps. Last
whorl is with prominent spiral ribs basally and weak to obsolete spiral ribs
or ribbons above. Ground color is white to pale pink. Last whorl is usually
with 2 broad yellowish brown spiral bands leaving 3 zones of reticulated
fine brown lines, at center, at base, and below shoulder. Brown areas are
interspersed with blackish brown axial lines and streaks. Shells are with
sparse tents and prominent. Shell size varies from 27 to 69 snails are predatory and venomous. They are capable of "stinging" humans and therefore
live ones should be handled carefully or not at all.

Toxin Type: Conopeptides; venom type not reported.

Conus ritae **(Petuch, 1995)**

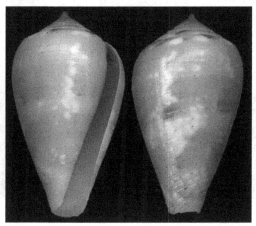

Apertural view   Abapertural view

Common Name: Not reported

Geographical Distribution: Endemic to Gorda Banks, off NE Honduras
(http://www.conchologistsofamerica.org/theshells/new_mar_g_e.asp-
Conchologists of America)

Habitat: At depths of 10–20 m

Identifying Features: Shell of this species is polished, glossy, inflated, bulbiform and widest just below shoulder. Spire is elevated, with distinctly convex whorls. Protoconch is elevated, projecting and mamillate. Color of the shell is deep red with two paler bands and entire body whorl is encircled with rows of tiny pale brown dots. Spiral whorls are with brown flammules. Maximum-recorded shell length is 27.5 mm. Like all species within the genus Conus, these snails are predatory and venomous. They are capable of "stinging" humans and therefore live ones should be handled carefully or not at all.

Toxin Type: Conopeptides; venom type not reported.

Conus rizali **(Olivera and Biggs, 2010)**

Apertural view   **Abapertural view**

Common Name: Not known

Geographical Distribution: Endemic to the Philippines

Habitat: Deep-water species; at depths between 100–200 m

Identifying Features: Shell of this species is medium-sized and biconic, with an unusually tall, straight and sharply pointed spire and a straight-sided body whorl. Shoulder is sharply angled. Outline is narrow. Larval shell has two whorls and this is followed by two teleoconch whorls that have a characteristic white-matte surface, which is somewhat crinkly. Starting with the fourth teleoconch whorl, there are 8–9 spotted spire whorls. Body whorl is characterized by shallow spiral ribbons with only a narrow interstitial space between them. These are broadly spotted in light yellow-brown. Shell size varies from 26 to 39 mm in length. Like all species within the genus Conus, these snails are predatory and venomous. They are capable of "stinging" humans and therefore live ones should be handled carefully or not at all.

Toxin Type: Conopeptides; venom type not reported.

Conus rosalindensis **(Petuch, 1998)** *(= Conus (Purpuriconus) rosalindensis)*

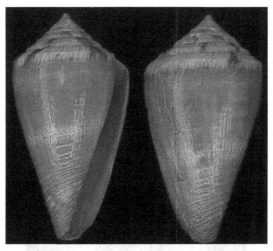

Apertural view   Abapertural view

Common Name: Not known

Geographical Distribution: Banks and islands off Honduras, Nicaragua and Jamaica

Habitat: Coral rubble between 3–15 m deep

Identifying Features: Shell of this species is smooth, with high polish. Spire is low and slightly stepped. Shoulder is sharply angled and coronated. Shell color is dark orange-tan with broad mid-body band. Band is overlaid with numerous, closely packed, hair-like, brown longitudinal flammules. Protoconch and early whorls are bright cherry-red and interior of aperture is pinkish purple. Maximum reported size of the shell is 26 mm. Like all species within the genus Conus, these snails are predatory and venomous. They are capable of "stinging" humans and therefore live ones should be handled carefully or not at all.

Toxin Type: Conopeptides; venom type not reported.

Conasprella rutila (Menke, 1843) (= *Conus rutilus*)

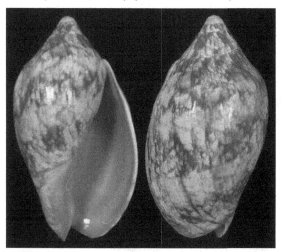

Apertural view   Abapertural view

Common Name: Fiery-red cone, fiery cone

Geographical Distribution: Endemic to Australia; from the south of New South Wales to Perth in Western Australia; along the north coast of Tasmania

Habitat: Intertidal and uppermost subtidal; on coral reef, in sand and short weed; depths from 1 m to 220 m

Identifying Features: This species has a very small and light shell. Last whorl is conical or ventricosely conical to broadly conical or broadly and ventricosely conical. Outline is convex to straight and straighter toward base. Left side is usually concave near base. Shoulder is subangulate to broadly carinate and weakly tuberculate to undulate. Spire is of low to moderate height and outline is concave to straight. Larval shell is of 1.25–1.5 whorls with a maximum diameter 0.7–0.9 mm. Postnuclear spire whorls are distinctly tuberculate to undulate. Teleoconch sutural ramps are concave and rarely with obsolete spiral striae. Last whorl is with weak spiral ribs or narrow ribbons near base. Color of the shell is from white or gray through yellow, pink or orange to violet and brown. Last whorl is usually with spiral rows of often alternating white and dark dots or dashes as well as with dark and white wavy to zigzag-shaped axial lines and streaks. In dark shells, last whorl may be encircled with a light band below center. Light shells are with 2–4 interrupted to solid dark bands. Larval whorls are white to dark brown. Postnuclear sutural ramps are white to gray or matching last whorl in color and immaculate or with brown radial markings. Aperture is matching the exterior surface in color. Once mature this species can reach a size ranging from 8 to 13 mm. Like all species within the genus Conasprella, these cone snails are predatory and venomous. They are capable of "stinging" humans and therefore live ones should be handled carefully or not at all.

Toxin Type: Conopeptides; venom type not reported.

Conus salzmanni **(Raybaudi and Rolan, 1997)**

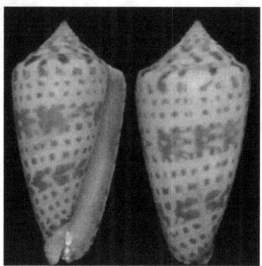

Apertural view   Abapertural view

Common Name: Not reported

Geographical Distribution: Gulf of Aden off the coast of Somalia and Djibouti; Yemen

Habitat: In depths from shallow water (Yemen population) to 150 m

Identifying Features: Protoconch of this species is with 1.75 to 2 whorls and remaining whorls are smooth. Adults of the species will grow to approx. 39 mm. It is a fish hunting species. Like all species within the genus Conus, these snails are predatory and venomous. They are capable of "stinging" humans and therefore live ones should be handled carefully or not at all.

Toxin Type: Conopeptides; venom type not reported.

Conus sanderi **(Wils and Moolenbeek, 1979)** *(= Conus (Leptoconus) paschalli, Dauciconus sanderi)*

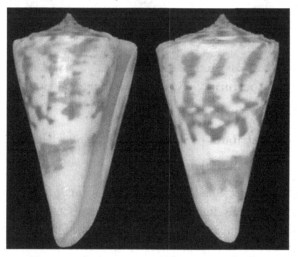

Apertural view   **Abapertural view**

Common Name: Sander's cone

Geographical Distribution: Endemic to Barbados

Habitat: At depths between 100 to 300 m, in muddy sand; Brazil

Identifying Features: Shell of this species is with stocky body whorl, wide shoulder, and proportionally low spire. Shell size varies from 24 to 57 mm. Average size of the shell is 24 mm. Like all species within the genus Conus, these snails are predatory and venomous. They are capable of "stinging" humans and therefore live ones should be handled carefully or not at all.

Toxin Type: Conopeptides; venom type not reported.

Conus scalptus **(Reeve, 1843)** (= *Asprella scalptus*)

Apertural view   Abapertural view

Common Name: Amadis cone

Geographical Distribution: Sulu Sea (*S.Philippines*), Papua New Guinea

Habitat: At depths of 0–20 m

Identifying Features: Shell of this species is small to moderately small and light to moderately light. Last whorl is ventricosely conical. Outline is convex and less so abapically. Aperture is somewhat wider at base than near shoulder. Shoulder is subangulate to rounded. Spire is of moderate height, and outline is concave to sigmoid. Larval shell is probably of 2 whorls with a maximum diameter 0.7–0.8 mm. Teleoconch sutural ramps are flat to slightly convex, with 1–2 increasing to 3–4 spiral grooves. Last whorl is with spiral grooves on basal third to half. Ribbons are grading to ribs anteriorly and weak in large specimens. Ground color is white. Last whorl is with rather closely spaced, yellow or reddish brown, interrupted and continuous spiral lines from base almost to shoulder and with sparse scattered small markings of the same color. Spiral pattern is of pale tan or orangish brown spiral bands on adapical half and within basal third. Shell size varies from 22 to 28 mm. Like all species within the genus Conus, these snails are predatory and venomous. They are capable of "stinging" humans and therefore live ones should be handled carefully or not at all.

Toxin Type: Conopeptides; venom type not reported.

Conus serranegrae **(Rolan, 1990)**

Apertural view   Abapertural view

Common Name: Sea snail

Geographical Distribution: Known only from the type locality: Serra Negra, Sal, Cape Verde

Habitat: Shallow water, attached under rocks or on rock platform with coarse sand under stones

Identifying Features: This species has a shell, which is perhaps broader than usual, with the spire slightly elevated, white with light brown blotches. A grayish white color predominates in the last whorl. A very fine-netted pattern is present over the last whorl, formed by zig-zag-shaped blotches of a light gray or pale grayish green color on a white background, arranged in bands as usual with an almost white patternless central band. Size of this small-sized species ranges between 10 and 20 mm. Like all species within the genus Conus, these snails are predatory and venomous. They are capable of "stinging" humans and therefore live ones should be handled carefully or not at all.

Toxin Type: Conopeptides; venom type not reported.

Conus sertacinctus (Röckel, 1986) (= *Asprella sertacinctus*)

Apertural view   Abapertural view

Common Name: Not known

Geographical Distribution: Endemic to the Solomon Islands; Southern India, Philippines and Indonesia; New Britain (Papua New Guinea) and Marshall Islands

Habitat: From shallow depths to 100 m on sand; fine white coral sand

Identifying Features: It is a small to medium sized shell (25–42 mm) which is solid although light weight. Spire is medium in height and straight to slightly concave in outline. Like all species within the genus Conus, these snails are predatory and venomous. They are capable of "stinging" humans and therefore live ones should be handled carefully or not at all.

Toxin Type: Conopeptides; venom type not reported.

Conus shikamai (Coomans, Moolenbeek and Wils, 1985) (= *Kioconus shikamai*)

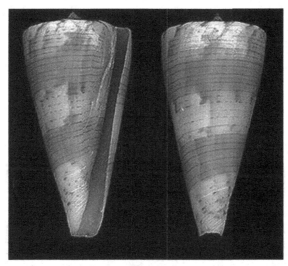

Apertural view **Abapertural view**

Common Name: Shikama's cone

Geographical Distribution: From Taiwan in the north through to Sulawesi (Indonesia) in the south including Philippines

Habitat: Between 100 and 240 m; sand and/or mud

Identifying Features: Shell of this species is medium-sized to moderately large and moderately solid-to-solid. Last whorl is usually conical. Outline is slightly convex below shoulder, then straight. Shoulder is angulate and irregularly undulate to weakly tuberculate. Spire is low and outline is concave. Larval shell is of 3–3.5 whorls with a maximum diameter 0.8–0.9 mm. First 3 postnuclear whorls are weakly tuberculate to irregularly undulate and adjacent whorls are nearly smooth. Late whorls are undulate. Teleoconch sutural ramps are slightly concave, with 2–3 increasing to 4–6 spiral grooves, obsolete in latest whorls. Last whorl is with variably spaced spiral ribs at base. Ground color is pale bluish violet. Last whorl is with a broad, usually continuous, brown spiral band on each side of center and with brown axial flames, usually crossing spiral bands and adjacent ground-color areas. Rather evenly spaced, fine dark brown spiral lines cover entire last whorl of subadult specimens but are mainly restricted to the brown spiral bands in adult shells. Larval whorls are brown. Teleoconch sutural ramps are heavily maculated with dark brown radial markings. Aperture is violet. Shell size varies from 45 to 70 mm. Like all species within the genus Conus, these snails are predatory and venomous. They are capable of "stinging" humans and therefore live ones should be handled carefully or not at all.

Toxin Type: Conopeptides; venom type not reported.

Conus stanfieldi **(Petuch, 1998) (= *Conus (Purpuriconus) stanfieldi*)**

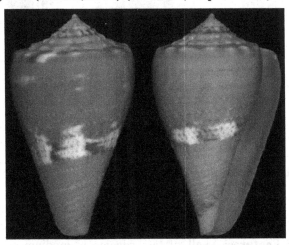

Abapertural view   Apertural view

Common Name: Stanfieldi cone

Geographical Distribution: North Atlantic Ocean; Type Locality: Northern Great Bahama Bank, Bahamas

Habitat: Reef wall; depth of 21 m

Identifying Features: Shell of this species grows up to 33 mm and is inflated, broad across the shoulder and smooth with high polish. Body is faintly sculptured with 8–10 slightly impressed spiral threads. Shell color is bright orange with wide, amorphous mid-body band of whitish-pink and scattered whitish-pink patches below shoulder. Like all species within the genus Conus, these snails are predatory and venomous. They possess a harpoon-like structure that allows them to stabbing and paralyze their prey.

Toxin Type: Conopeptides; venom type not reported.

Conus stocki **(Coomans and Moolenbeek, 1990)**

Abapertural view   Apertural view

Common Name: Not reported

Geographical Distribution: Off the coast of Oman at Masirah Island and Masqat

Habitat: At depths from shallow water to 25 m among stones

Identifying Features: Shell of this species is elongate, obconical, thin, glossy. Maximum length of the shell is 45 mm, width 16.3 mm and height of spire 16.2 mm. Body whorl is mostly straight in profile view and a little convex near the shoulder. Lower part is grooved. Aperture is narrow. Spire is concave, consisting of 9 whorls. Apical angle is about 100°. Main color is bluish-white with brown blotches and punctuated brown-white spirals. Apical whorls are with irregular brown spots above the suture and on the last whorl there are ten dark brown spots. This species has a worm hunting type radula. Like all species within the genus Conus, these snails are predatory and venomous. They are capable of "stinging" humans and therefore live ones should be handled carefully or not at all.

Toxin Type: Conopeptides; venom type not reported.

Conus stramineus (Lamarck, 1810) (= *Asprella stramineus*)

Apertural view   Abapertural view

Common Name: Subulate cone

Geographical Distribution: Restricted to Indonesia; Moluccas and off the South West coast of Java

Habitat: Shallow waters

Identifying Features: It is a medium to large sized (30–50 mm in length) conical shell. Shoulder is subangulate and smooth. Body whorl is almost straight in outline and is only slightly curved in towards the shoulder. It is shiny and cream to off-white with 12–14 spiral rows of squarish brown spots and blotches. Like all species within the genus Conus, these snails are predatory and venomous. They are capable of "stinging" humans and therefore live ones should be handled carefully or not at all.

Toxin Type: Conopeptides; venom type not reported.

Conus sugimotonis (**Kuroda, 1928**) (= *Kioconus sugimotonis*)

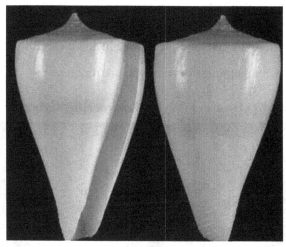

Apertural view   Abapertural view

Common Name: Sea snail

Geographical Distribution: There are two populations: in the north this species occurs from Japan to the Philippines including Taiwan; it also occurs in south Queensland, Australia and the Coral Sea

Habitat: At depths of between 200 and 400 m probably in sand or mud

Identifying Features: Shell of this species is moderately large to large and usually solid. Last whorl is conical to ventricosely conical or slightly pyriform. Outline is convex at adapical fourth to half, straight or slightly concave below. Shoulder is angulate. Spire is low. Outline is concave to sigmoid and is often with early whorls projecting from an otherwise flat or slightly domed spire. Larval shell is of about 3 whorls with a maximum diameter about 1 mm. First 3–5 postnuclear whorls are tuberculate. Teleoconch sutural ramps are flat to slightly concave, with 1 increasing to 5–8 spiral grooves, sometimes with additional spiral striae in latest whorls. Last whorl is with weak to distinct spiral ribs and ribbons at base. Ground color is white. Last whorl is immaculate or variably shaded with yellow or tan. Larval whorls are white or beige. Early teleoconch sutural ramps may be tinged with yellow. Outer margins of sutural ramps are sometimes with a varying number of brown dots, persisting at edge of shoulder in some specimens. Aperture is white, sometimes bluish white. Shell of this species has a size range of 60–120 mm. Like all species within the genus Conus, these snails are predatory and venomous. They are capable of "stinging" humans and therefore live ones should be handled carefully or not at all.

Toxin Type: Conopeptides; venom type not reported.

Conus suratensis **(Hwass, 1792)** *(= Dendroconus suratensis)*

Apertural view   Abapertural view

Common Name: Surat cone

Geographical Distribution: Indo-West Pacific, from East Africa, southern India and Sri Lanka to Melanesia; north to the Philippines and south to Indonesia

Habitat: Littoral and shallow sublittoral zones to a depth of about 20 m; sandy and muddy substrata

Identifying Features: Shell of this species is large, thick and heavy and inflated, with a very low spire and slightly elevated apex. Spire whorls are flat to slightly concave, without distinct sculpture. Shoulder is broad and rounded and flat to shallowly concave below suture. Body whorl is moderately glossy, nearly smooth except for several spiral grooves anteriorly and a rather broad and low spiral ridge on columella. Color: outside of shell is cream to white, with numerous, small dark brown spots on body whorl, reduced to narrow, transverse dashes and arranged in irregular axial bands. Dark spots sometimes are more or less fused into short axial flammules towards the shoulder. Anterior end of shell is broadly tinged bright orange or light fawn. Spire whorls are white, often with orange hue, and with curved, blackish brown radiating lines. Aperture is entirely white inside. Maximum shell length is 15 cm. Like all species within the genus Conus, these snails are predatory and venomous. They are capable of "stinging" humans and therefore live ones should be handled carefully or not at all.

Toxin Type: Conopeptides; venom type not reported.

Conus sutanorcum **(Moolenbeek, Röckel and Bouchet, 2008) (=** *Phasmoconus sutanorcum***)**

Apertural view   **Abapertural view**

Common Name: Not known

Geographical Distribution: Off the coast of the island of Fiji, primarily around Viti Levu

Habitat: At depths between 32 and 50 m; soft bottom substrate offshore areas

Identifying Features: Shell of this species is medium-sized and moderately solid. Shape is conical to broadly ovate. Spire is concave.

Protoconch is glassy white with two nearly smooth tapering whorls. Teleoconch is of 9.5 whorls. First 3 post nuclear whorls are with fine nodules gradually disappearing. Spire whorls are starting with one and ending with 4 spiral grooves. Suture is rather deep. Color is white with irregular axial brown markings, penultimate whorl with about 12 markings. Last whorl is with about 35 spiral cords, which are narrower than the axially striated interspaces. On the basal part of the whorl the interspaces are about 5 times as broad as the cords. Upper cords are brown with now and then a white area in between, on lower cords the brown and white portions are nearly equal. Base is white. Aperture is rather slender and white with the brown lines shining through the edge. Dimensions: height 29.7 mm, width 13.8 mm. Like all species within the genus Conus, these snails are predatory and venomous. They are capable of "stinging" humans and therefore live ones should be handled carefully or not at all.

Toxin Type: Conopeptides; venom type not reported.

Conus suturatus **(Reeve, 1844)** (= *Lithoconus suturatus*)

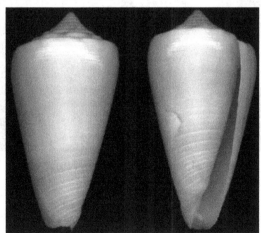

Abapertural view   Apertural view

Common Name: The sutured cone

Geographical Distribution: From Papua New Guinea to NW Australia and Queensland, with a subspecies in Hawaii; Australian waters; Philippines

Habitat: Rare on exposed intertidal parts of coral reef, more frequent in subtidal habitats in 7–150 m; often on muddy bottoms

Identifying Features: Shell of this species is moderately small to medium-sized and moderately solid-to-solid. Last whorl is conical and ventricosely conical or broadly conical. Outline is straight or variably convex adapically. Shoulder is subangulate to angulate. Spire is low and outline is concave to straight. Larval shell is multispiral with a maximum diameter about 0.7 mm. First 5–8 postnuclear whorls are tuberculate. Teleoconch sutural ramps are with a pronounced subsutural ridge and a distinct ribbon between 2 spiral grooves. In large specimens, last ramp is sometimes with 3–4 spiral grooves. Last whorl is with variably spaced deep spiral grooves at basal fourth to third. Ground color is white and sometimes suffused with pink or violet. Last whorl is usually with 3 orange or pink spiral bands, below shoulder and on both sides of center. Adapical band is usually pale. Sometimes, additional variously sized spiral rows are seen. Squarish, yellowish brown spots form clusters overlying color bands. Rows vary in number and arrangement and sometimes contain white spots. Base and siphonal fasciole are light violet. Larval whorls are white. Teleoconch sutural ramps are immaculate or with orangish brown radial markings producing separated spots or bars along shoulder edge. Aperture is white to violet. Shell size ranges from 30 to 43 mm. Like all species within the genus Conus, these snails are predatory and venomous. They are capable of "stinging" humans and therefore live ones should be handled carefully or not at all.

Toxin Type: Conopeptides; venom type not reported.

Conus telatus **(Reeve, 1848)** *(= Cylinder telatus)*

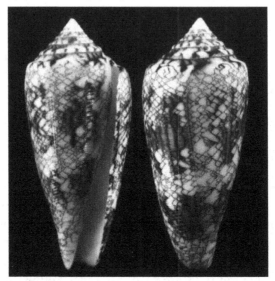

Apertural view   **Abapertural view**

Common Name: Webbed cone

Geographical Distribution: Off Palawan, Balabac and Cuyo in the Philippines

Habitat: In 2–100 m; on coral reef, on sand substrate among rocks and coral blocks

Identifying Features: Shell of this species is medium-sized to large and moderately solid-to-solid. Last whorl is ventricosely conical to conoid-cylindrical, also sometimes conical in typical ("smooth") form. Outline is convex to nearly straight and is often more convex in strongly sculptured shells (form rugosus) than in typical shells. Left side is slightly concave at basal third in typical form, to strongly concave in form rugosus. Aperture is somewhat wider at base than near shoulder. Shoulder is angulate and smooth in typical form. It is weakly tuberculate in form rugosus except for large adults. Spire is of moderate height and outline is concave to nearly straight, usually less concave in typical form. Larval shell is of about 2 whorls with a maximum diameter 1–1.2 mm. In typical form, first 5–7 postnuclear whorls are rather weakly tuberculate and in form rugosus, first 8–9.5 postnuclear whorls are tuberculate. Teleoconch sutural ramps are concave. Last whorl is with weak spiral ribs on basal third in typical form, with closely spaced and finely granulose ribs from base to shoulder in form rugosus. Ground color is white. In typical form, last whorl is with fine reticulated lines edging very small to medium-sized tents. On each side of center, yellowish brown blotches are forming a spiral band interrupted by larger tents. Brown blotches are interspersed with variable dark brown axial lines. Larval whorls and a few adjacent postnuclear sutural ramps are immaculate white. Following ramps are matching last whorl in color pattern. Aperture is white, rarely suffused with cream. Size of the shell varies from 48 to 100 all species within the genus Conus, these snails are predatory and venomous. They are capable of "stinging" humans and therefore live ones should be handled carefully or not at all.

Toxin Type: Conopeptides; venom type not reported.

Conus terryni **(Tenorio and Poppe, 2004)**

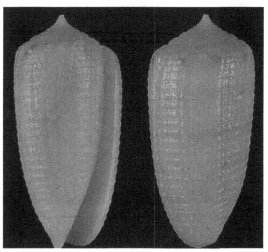

Apertural view   Abapertural view

Common Name: Not reported

Geographical Distribution: Of the coast of Aliguay Island in the Philippines

Habitat: At depths of approx. 60–150 m

Identifying Features: Shell of this species is moderately small and solid and heavy shell. Profile is very characteristic conoid-cylindrical to cylindrical with a moderate spire and a rounded shoulder. Outline of last whorl is convex at the adapical third to straight and parallel -sided. Outline of spire is straight with a very well marked suture. Protoconch is absent sometimes. Color of aperture is white and ground color of shell is ivory white. Shell size ranges from 25 to 31 mm. Like all species within the genus Conus, these snails are predatory and venomous. They are capable of "stinging" humans and therefore live ones should be handled carefully or not at all.

Toxin Type: Conopeptides; venom type not reported.

Conus tirardi **(Rockel and Moolenbeek, 1996) (=** *Rhizoconus tirardi*)

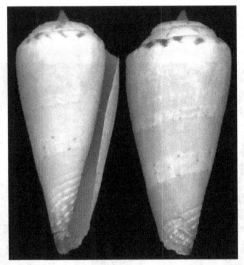

Apertural view   Abapertural view

Common Name: Not reported

Geographical Distribution: A new species from the Pacific Ocean; New Caledonia around Nouméa, Eastern Coral Sea and Off Pitcairn Island

Habitat: At depths ranging from intertidal to 20 m

Identifying Features: Shell of this species is light, with a multispiral protoconch. Outline is with a slightly convex apex and almost straight below. Color is light beige, with 3 spiral bands on last whorl. Base is purple with white spiral cords and irregular fine brown spots. Maximum reported size of the shell is 31 all species within the genus Conus, these snails are predatory and venomous. They are capable of "stinging" humans and therefore live ones should be handled carefully or not at all.

Toxin Type: Conopeptides; venom type not reported.

Conus tisii **(Lan, 1978) (=***Asprella tisii, Conus (Rhizoconus) tisii***)**

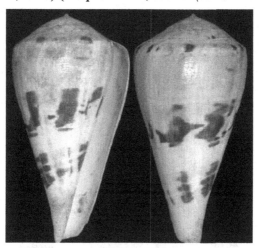

Apertural view   Abapertural view

Common Name: Not known

Geographical Distribution: Indo-Pacific; Taiwan, Philippines

Habitat: In 120–400 m

Identifying Features: Shell of this species is large and solid to heavy. Last whorl is conical or ventricosely conical. Outline is convex adapically and straight (right side) or concave (left side) below. Shoulder is subangulate to rounded. Spire is of low to moderate height and outline is sigmoid. First postnuclear whorls are domed. Larval shell is of 2–2.5 whorls with a maximum diameter 1.1–1.2 mm. First 5–6 postnuclear whorls are tuberculate. Teleoconch sutural ramps are flat, with 2 increasing to 7–9 spiral grooves. Last whorl is with numerous closely set spiral ribs at base, followed by spiral striae to shoulder. Ground color is white to pale violet. Last whorl is with

2 darker violet spiral bands bearing brown axial blotches, on basal third and just above center. Adult specimens are also with irregular brown spots and axial streaks. Shoulder edge is with brown spots, occasionally also present in preceding whorls. Aperture is white to light purple. Size of the shell ranges from 98 to 154 mm. Like all species within the genus Conus, these snails are predatory and venomous. They are capable of "stinging" humans and therefore live ones should be handled carefully or not at all.

Toxin Type: Conopeptides; venom type not reported.

Conus tribblei (Walls, 1977)

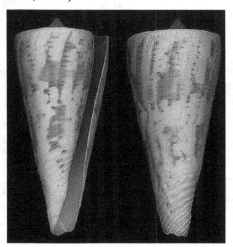

Apertural view   Abapertural view

Common Name: Tribble's cone

Geographical Distribution: From Japan to Philippines including Taiwan; north-west Australia and Solomon Islands; subspecies *C. t. queenslandis* in New Caledonia and Queensland

Habitat: In 100–300 m; sand and/or mud

Identifying Features: Shell of this species is moderately large to large and usually solid to moderately heavy. Last whorl is narrowly conical-to-conical and occasionally ventricosely conical. Outline is slightly convex at adapical fourth or sometimes two-thirds, straight below, often with a slightly concave central area. Shoulder is angulate to broadly carinate. Spire is low and outline is concave. Early whorls are usually projecting from an otherwise rather flat spire. Larval shell is of 3–3.5 whorls with a maximum diameter 0.8–1 mm. First 3–6 postnuclear whorls are tuberculate. Late whorls are usually carinate. Teleoconch sutural ramps are flat to slightly concave, with 2 increasing to 4–7 spiral grooves, often weaker in latest whorls. Shells are with variably spaced and strongly granulose spiral ribs from base to sub-shoulder area intergrade with shells with variably prominent, smooth or granulose ribs at base. Ground color is white, often suffused with cream in *C. t. queenslandis*. Last whorl is with tan to brown axial streaks and blotches on adapical two-thirds and a continuous or interrupted spiral band of the same color on each side of center. Anterior color band is often weaker or even absent. *C. t. queenslandis* with cream to orangish brown color bands and without axial streaks and blotches. Coarse dashed or dotted brown spiral lines may be present but vary in number and arrangement. Base is white or sometimes pale yellow. Shell size ranges from 60 to 138 mm. These snails are predatory and venomous. They are capable of "stinging" humans and therefore live ones should be handled carefully or not at all.

Toxin Type: Conopeptides; venom type not reported.

Conus tuticorinensis **(Röckel and Korn, 1990)** (= *Quasiconus tuticorinensis*)

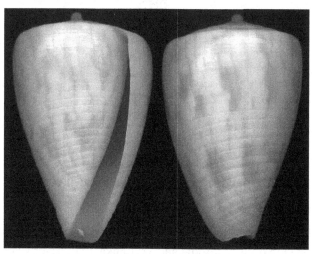

Apertural view   **Abapertural view**

Common Name: Not known

Geographical Distribution: India, Sri Lanka, Philippines and New Guinea

Habitat: The exact depth of occurrence and nature of bottom is not known

Identifying Features: Shell of this species is small and moderately solid. Body whorl is broadly conical and outline is convex adapically. Shoulder is angulate. Spire is low and outline is straight. Body whorl is with flat spiral ribbons from base to shoulder. Ground color is white, suffused with pale violet on last whorl. Body whorl is with orange-brown "brick wall pattern" of about 12 spiral lines and irregular axial lines. Similarly colored irregular flecks are spirally aligned below shoulder and each side of center. Apex is white. Aperture is reddish white. Shell size ranges form 22 to 30 snails are predatory and venomous. They are capable of "stinging" humans and therefore live ones should be handled carefully or not at all.

Toxin Type: Conopeptides; venom type not reported.

Conus vimineus **(Reeve, 1849) (=** *Conasprella viminea***)**

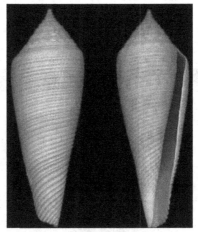

Abapertural view   Apertural view

Common Name: Wickerwork cone

Geographical Distribution: Widely distributed in three separate population groups: (1) South-east India and Sri Lanka, (2) Western Thailand, and (3) The Philippines

Habitat: From 75 to 280 m depth on fine mud bottoms

Identifying Features: Shell of this species is moderately small, light and fragile and elongate. Body whorl is narrowly conical. Outline is convex adapically, and straight below. Entire body whorl is with deep, axially striate spiral grooves separating regularly spaced ribs near base and variably arranged ribs and narrow ribbons above. Elevations of ribs are smooth. Shoulder is broadly rounded and is not distinct from spire. Spire is moderately high and is sharply pointed with straight sides. Aperture is uniformly narrow. Outer lip is very thin and is strongly sloping below level of shoulder and straight. Ground color is pale brown. Body whorl is with spiral rows of rectangular light brown spots on ribs, fusing into axial streaks that cluster in three indistinct spiral bands below shoulder, above center and near base. Spire is pale brown. Aperture is white. Adults of the species will grow to approx. 43 mm. Like all species within the genus Conasprella, these cone snails are predatory and venomous. They are capable of "stinging" humans and therefore live ones should be handled carefully or not at all.

Toxin Type: Conopeptides; venom type not reported.

Conus violaceus **(Gmelin, 1791)**

Apertural view   Axial view   Abapertural view

Common Name: Violet cone

Geographical Distribution: Eastern Indian Ocean including the coast of East Africa, and Madagascar and the surrounding islands

Habitat: Shallow water, from 3 to 12 m; coral rubble, among seaweed, and in algae encrusted crevices

Identifying Features: Shell of this species is medium sized, moderately solid and glossy. Body whorl is narrowly cylindrical. Outline is convex over adapical third and straight below to almost uniformly straight and parallel-sided. Shoulder is indistinct from spire. Spire is of moderate height and outline is convex. Sutures are narrowly channeled. Spire tip is rounded and sharp. Body whorl is with widely separated narrow spiral ribs from base to shoulder. Aperture is narrow and wider anteriorly. Outer lip is thick, straight and is sloping below shoulder. Ground color is white. Body whorl is with light brown axial streaks and with three dark-brown spiral bands above center, within basal third and at shoulder. Spire is light-brown with early whorls which are white. Base is violet. Aperture is white and outer color is visible along the inner lip margin. Typical *size* for shells of this species is between 50–93 mm in *length*. Like all species within the genus Conus, these snails are predatory and venomous. They are capable of "stinging" humans and therefore live ones should be handled carefully or not at all.

Toxin Type: Conopeptides; venom type not reported.

Conus voluminalis **(Reeve, 1843) (= *Conus macara*)**

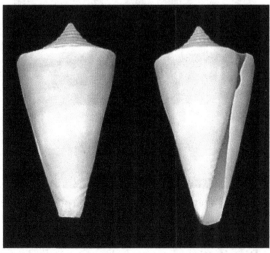

Abapertural view   Apertural view

Common Name: Sea snail

Geographical Distribution: Indo-Pacific-ranging from Maldives to West of Australia and north to Ryuku Island; coast of Taiwan, Solomon Islands and Papua New Guinea

Habitat: At intertidal depths to 150 m; coral and rocky shorelines

Identifying Features: Shell of this species is medium-sized to moderately large and moderately solid-to-solid. Last whorl is conical. Outline is almost straight and slightly convex adapically in small specimens. Shoulder is angulate to broadly carinate and is often outwardly curved producing a concave outline at subshoulder area. Spire is of low to moderate height and outline is concave to slightly sigmoid. Apex may project from an otherwise almost flat spire. Larval shell is of 2–2.25 whorls with a maximum diameter 0.9–1 mm. First 4–6 postnuclear whorls are weakly tuberculate. Teleoconch sutural ramps are flat to slightly concave, with 1–2 increasing to 4–5 spiral grooves, obsolete in late whorls. Last whorl is with a few spiral ribs at base. Ground color is white, yellow, orange or pale violet. Last whorl is with a broad, continuous to sometimes interrupted, yellow or orange to brown spiral band on each side of center, sometimes extending to shoulder and base. Long light to dark brown axial streaks may extend from shoulder ramp to base. Dotted to continuous spiral lines may cover entire last whorl, but vary in number and arrangement. Shells are with no axial streaks or spiral lines occur. Larval whorls are white to beige. Early teleoconch sutural ramps are immaculate and are often light pink. Late ramps are either of immaculate ground color or more frequently with orange to dark brown radial markings. Aperture is white, yellow, orange or violet. Adults of this species typically grow between 40 to 72 mm. Like all species within the genus Conus, these snails are predatory and venomous. They are capable of "stinging" humans and therefore live ones should be handled carefully or not at all.

Toxin Type: Conopeptides; venom type not reported.

Conus vulcanus **(Tenorio and Afonso, 2004)**

Apertural view   Abapertural view

Common Name: Not reported

Geographical Distribution: Baía das Gatas and Derrubado; extending the range of this species to the north of Boavista; Porto Ferreira

Habitat: Rocky reef; large rock platforms usually covered with algae and some between 1 and 3 m depth

Identifying Features: Shell of this species is small to moderately small. Profile of the shell is conical or ventricosely conical, with a low to moderate spire and a rounded shoulder. Outline of the last whorl is straight or most often convex, rather concave abapically resulting in a slightly pyriform shape. Spire is almost always eroded and outline is concave in those specimens in which it is preserved. Teleoconch sutural ramps are flat, with spiral grooves. Suture is well marked. Last whorl is smooth, with variably spaced fine spiral ribs present in the basal quarter. Aperture is narrow and wider towards the base. Shell has a very dark brown color, almost black in occasions, which appears lighter ventrally. Very fine characteristic dark brown, equally spaced spiral lines are usually visible, specially on the ventral side and in young specimens. There is a white spiral band around the middle portion of the last whorl, which is covered by a dark brown reticulated pattern. The region between the shoulder and the midbody band displays often a fine reticulated pattern of white dots, although in many occasions it appears solid dark brown. Basal third is usually void of any reticulated pattern, being dark brown. On the shoulder there is often present a fine spiral band of a lighter brown color. Just above the shoulder and in the late sutural ramps of the spire, the reticulated pattern is replaced by irregular dark brown blotches on white. Color of the aperture is white in the larger specimens. In smaller specimens, the aperture is stained with purple-brown, with the inner lip, which is white, and two white bands, one near the shoulder and another below the midbody are seen. Aperture is white within. Columella is white. Size of the shell is 22.5 x 14.5 mm. Like all species within the genus Conus, these snails are predatory and venomous. They are capable of "stinging" humans and therefore live ones should be handled carefully or not at all.

Toxin Type: Conopeptides; venom type not reported.

Conus wilsi **(Delsaerdt, 1998)**

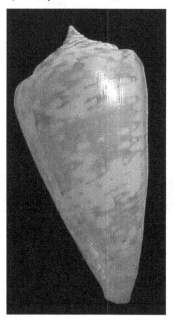

Abapertural view

Common Name: Not reported

Geographical Distribution: A new species from the Red Sea

Habitat: Currently no information available on the habitat and ecology of the species, including depth information

Identifying Features: Last whorl of the shell of this species is with a straight and smooth outline. Spire is almost flat. Sutural ramps are with 3 spiral grooves. Patterns of brown spots or veins are seen on a white to violet-white background. Central white band is edged by spiral row of brown spots or flames. Maximum size of the shell is 30 mm. Like all species within the genus Conus, these snails are predatory and venomous. They are capable of "stinging" humans and therefore live ones should be handled carefully or not at all. Type locality: South of Queasier, Red Sea, Egypt.

Toxin Type: Conopeptides; venom type not reported.

Conus worki **(Petuch, 1998) (**= *Conus (Dauciconus) worki*)

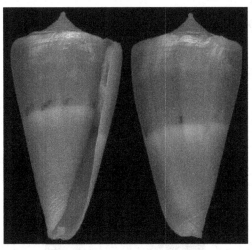

Apertural view   **Abapertural view**

Common Name: Sea snail

Geographical Distribution: Endemic to Brazil where it is found from Rio Grande do Norte State to Cabo Frio, Rio de Janeiro State

Habitat: At depths between 10 and 20 m in coral sand

Identifying Features: This new species is most similar to *C.daucus*, but differs in consistently being narrower, more slender; with distinctly more channeled spire whorls which are bordered by a raised carina. It is paler than *C.daucus* in coloration, with a white mid-body band that is bounded by brown flammules and dots, and "most importantly," it has a yellow protoconch instead of pink. Adults grow to approximately 35 mm. Like all species within the genus Conus, these snails are predatory and venomous. They are capable of "stinging" humans and therefore live ones should be handled carefully or not at all.

Toxin Type: Conopeptides; venom type not reported.

Conus zylmanae **(Petuch, 1998)** (= *Conus (Magelliconus) zylmanae*)

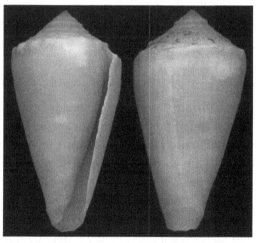

Apertural view   Abapertural view

Common Name: Sea snail

Geographical Distribution: Endemic to the Bahamas where it is found off New Providence

Habitat: At depths of 1–7 m; coral and coral rubble

Identifying Features: Shell of this species is more pyriform. Spire is high with more numerous and smaller coronations Shell color is khaki-green, with smaller white flammules. Maximum-recorded shell length is 24 mm. Like all species within the genus Conus, these snails are predatory and venomous. They are capable of "stinging" humans and therefore live ones should be handled carefully or not at all.

Toxin Type: Conopeptides; venom type not reported.

3.2   TURRID SNAILS (FAMILY: TURRIDAE)

The turrids of the family Turridae are predatory sea snails (marine gastropod mollusks), which vary in size from minute to moderately large. With more than 4,000 species, the Turridae is the largest mollusk family and group of marine caenogastropods. Turrids are found worldwide in every sea and ocean from both poles to the tropics. They occur from the low-intertidal zone to depths of more than 8,000 m. However, most species of turrids are found in the neritic zone. The biology, feeding ecology and phylogenetic relationships of marine snails in the family Turridae remain poorly understood.

3.2.1   SHELL DESCRIPTION

The adult shells of different species of turrids are between 0.3 and 11.4 cm in height. Shape of the shells is more or less fusiform, varying from very high-spired to broadly ovate. Whorls are elongate to broadly conical. Sculpture is very variable in form, but most have axial sculpture or spiral sculpture (or a combination of both). Others may be reticulate, beaded, nodulose, or striate. Aperture of the shell very often has a V-shaped sinus or notch, an indentation on the upper end of the outer lip. This accommodates the anal siphonal notch, commonly known as the "turrid notch." Siphonal canal is usually open, varying from short and stocky to long and slender. Position of the turrid notch of the shell and the form and sculpture of the whorls has been the primary methods of classifying the turrids. Columella is usually smooth and operculum is horny, but is not always present. Turrids are carnivorous, predatory gastropods. Most species have a poison gland used with the toxoglossan radula, used to prey on vertebrates and invertebrate animals (mostly polychaete worms) or in self-defense. Some turrids have lost the radula and the poison gland. The radula, when present, has two or three teeth in a row. It lacks lateral teeth and the marginal teeth. Female turrids lay their eggs in lens-shaped capsules.

Turridrupa cerithina **(Anton, 1838) (= *Crassispira cerithina*)**

Abapertural view   Apertural view

Class: Gastropoda; **Subclass**: Caenogastropoda

Order: Neogastropoda; **Superfamily**: Conoidea

Family: Turridae

Common Name: Sea snail

Geographical Distribution: Tropical Indo-Pacific.
Kii Peninsula south, tropical Indo-Pacific Ocean region.

Habitat: Intertidal to 10 m, gravel bottoms.
Gravel bottom of the intertidal zone −10 m.

Identifying Features: Shell of this species is reddish brown with pale granules in column. There is no characteristic turrid shape in this species. But it has a notch or slip on the outside of the lip of the shell. Shell size varies between 15 mm and 24 mm. It feeds on a variety of foods either by engulfing or stinging their prey. As these snails are venomous, they should be carefully handled.

Toxin Type: P-like crassipeptide superfamily, cce9a and cce9b (Imperial et al., 2014; Cabang et al., 2011) (http://www.uniprot.org/uniprot/G8FZS4; Kastin, 2013). This peptide has caused lethargy in mice (King, 2015).

Gemmula disjuncta **(Laws, 1936)**
   Image not available

Common Name: Not reported

Geographical Distribution: New Zealand

Habitat: No data

Identifying Features: Shell of this species is small and its spire is about twice height of aperture. Protoconch is not observed. Post-embryonic whorls are about 7 in number, keeled at or slightly below. Keel is ornamented by a row of blunt, heavy, rounded tubercles, which are about 14 in number on penultimate whorl, but becoming obsolete on the body. On the keel and in intervals between tubercles there are several fine threads. Subsutural border has subobsolete nodules and is weakly chordate. Between sutural border and keel there are several fine threads. Base carries distinct, regular cords, which are spaced evenly over earlier part, but becoming closer and finer on neck. Outer lip is with a deep sinus on keel. Inner lip is callused and a callus thickening is on neck. Height and width of the shell are 11.5 mm and 4.2 mm, respectively. As it is a newly described species, nothing much is known about its biology and ecology.

Toxin Type: Turripeptides: venom type not reported.

Gemmula speciosa (Reeve, 1843)

Apertural view   **Abapertural view**

Common Name: Splendid gem-turris

Geographical Distribution: Indo-W Pacific

Habitat: Deep-water species

Identifying Features: Shell of this species is fusiform, with tall spire, and long, tapered, and slightly flexed anterior canal. Whorls are angulate and carinate at just below middle whorl height. Base is rather suddenly contracted. Primary spirals are plain, thin but sharply raised. Secondary sculpture consists of from one to three plain weak threads in the interspaces of the primaries. Surface is crowded with weak, crisp axial threads. Color pattern is of light brown to darker golden brown spirals on a buff background. Peripheral carina is uniformly colored, and all of the primary spirals are similarly tinted light brown to golden brown. In typical specimens, there are no interrupted markings, dots or dashes. The light amber protoconch is typical of Gemmula, and is followed by nine teleoconch whorls in adult shells. In most specimens, the brown primary spirals become obsolete in the anterior siphonal canal, which becomes pure white. Adult shell of this species is 50–78 mm in height. Because of the deep habitats, little is known about the feeding habits of this species. These snails are venomous with disulfide-rich polypeptides in their venom ducts. These bioactive peptides become a resource for novel pharmacologically active compounds.

Toxin Type: Turripeptides – P-like (three disulfides): Gsp9a, Gsp9.1, Gsp9b and Gsp9.2 (Kastin, 2013; Aguilar et al., 2009; King, 2015);

Turripeptide Gsp9.2 (http://www.uniprot.org/uniprot/P0C850);

Turripeptide Gsp9.1 (http://www.uniprot.org/uniprot/P0C845);

Turripeptide Gsp9.3 (http://www.uniprot.org/uniprot/P0DKT3);

Recombinant Gemmula speciosa Toxin Gsp9b (http://www.amazon.com/Recombinant-Gemmula-speciosa-Toxin-Protein/dp/B00X80YDLQ);

Recombinant Gemmula speciosa Toxin Gsp9.2 (http://www.amazon.com/Recombinant-Gemmula-speciosa-Gsp9–2-Protein/dp/B00X8K8OD4);

Recombinant Gemmula speciosa Turritoxin XIV-18 (http://www.mybiosource.com/prods/Recombinant-Protein/Turritoxin-XIV-18/datasheet.php?products_id=1244410);

Turritoxin NCR-01 Alternative name(s): Turritoxin NCR-02 (http://www.cusabio.com/Recombinant-Protein/Recombinant-Gemmula-speciosa–Turritoxin-NCR-01–613495.html).

Gemmula sikatunai **(Olivera, 2004)**

Apertural view

Common Name: Sea snail

Geographical Distribution: Western Central Pacific: Philippines

Habitat: Deeper tropical waters at depths between 50 and 500 m

Identifying Features: Shell of this species is fusiform, with tall spire and a long, relatively broad canal. Whorls are strongly angulate and strongly carinate. Peripheral carina is gemmate, with two brown cords at the margins, the more posterior being somewhat stronger. Particularly in the larger whorls, the center of the carina and gemmae are lighter colored than the bordering cords. On the body whorl, the two brown cords on the carina are well separated by the lighter-colored central area in mature specimens. Primary spirals are sharply raised. Anterior to the peripheral carina, there are usually six primary spirals on the body whorl and the base and are uniformly light brown to golden brown in color. There are 5–7 additional primary spirals on the canal, becoming progressively lighter and weaker in some specimens, but remaining quite strong almost to the end of the canal in other specimens. Secondary sculpture consists of from 1–3 weak white threads in the interspaces and these become strong towards the base and in the canal. Protoconch is polygyrate with axial ribs, typical of Gemmula, and there are ten teleoconch whorls in adult shells. Adult shell is 40–51 mm in height Because of the deep habitats, little is known about the feeding habits of this species. These snails are venomous with disulfide-rich polypeptides in their venom ducts. These bioactive peptides become a resource for novel pharmacologically active compounds.

Toxin Type: All the species of this genus are venomous and possess toxic turripeptides venom type not reported.

Gemmula lululimi (Olivera, 1999)

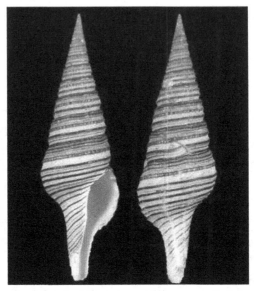

Apertural view   **Abapertural view**

Common Name: Gem turrid

Geographical Distribution: Philippines

Habitat: Deep-water habitat

Identifying Features: Spire of this species is high, cone-shaped. Adult specimens have 15–17 whorls. Spire angle is 25–27°. There are three primary spiral cords, continuous brown colored, in each spire whorl. Subsutural fold has a major primary cord, which is raised and in some larger specimens, a second cord is found immediately adjacent to the suture. There is a pronounced depression between the subsutural primary cord and the sinus cord. Sinus cord has gemmules, and in later whorls the two margins are brown, with the central gemmae lighter in color, almost white in some specimens. Sinus cord is not located peripherally, but is posterior to the periphery. A strong brown cord is present at the periphery in all the spire whorls. On the body whorl, there are 5–7 brown primary cords anterior to the sinus cord, with 4–6 brown primary cords on the anterior canal. The very tip of the canal is white, lacking the brown spiral cords in most (but not all) specimens. Specimens range in size from 37 to 90 mm.

Because of the deep habitats, little is known about the feeding habits of this species. These snails are venomous with disulfide-rich polypeptides in their venom ducts. These bioactive peptides become a resource for novel pharmacologically active compounds.

Toxin Type: All the species of this genus are venomous and possess toxic turripeptides; venom type not reported.

Gemmula kieneri **(Doumet, 1840)**

Apertural view   Abapertural view

Common Name: Kiener's gem-turris

Geographical Distribution: Japan, S. Africa and Australia

Habitat: Water depth of 50–346 m

Identifying Features: Shell of this species is robust, fusiform, with tall spire and long, rather straight anterior canal. Spire whorls are with a strong square-cut keel situated below the middle, not prominently projecting and sculptured, with closed spaced rectangular gemmules that are laterally compressed. There are regular squarish spots between the gemmules. The strong complex subsutural fold is irregularly blotched with brown, and usually consists of a primary cord (sometimes two) and additional threads.

Between the subsutural fold and the peripheral keel, there are usually 3–4 sharply raised, slightly imbricate threads. Between the primary keel and the lower suture, there is one primary cord and several threads. On the base, exclusive of the anterior canal, there are 6 primary cords, with irregularly disposed brown spots, with between 1–3 interstitial threads. Surface is covered with dense lamellate axial growth threads that imbricate the secondary spiral threads. Adult shell is up to 73 mm in height. Because of the deep habitats, little is known about the feeding habits of this species. These snails are venomous with disulfide-rich polypeptides in their venom ducts. These bioactive peptides become a resource for novel pharmacologically active compounds.

Toxin Type: Turripeptides – P-like (three disulfides): Gkn 91 (Kastin, 2013; Aguilar et al., 2009; King, 2015).

Gemmula sogodensis **(Olivera, 2004)**

Apertural view   Abapertural view

Common Name: Gem-turris

Geographical Distribution: Western Central Pacific: Philippines

Habitat: Deep-water species

Identifying Features: Adult shell of this species is with 11–13 whorls. Spire is relatively high with an angle of 34–36°. Anterior canal is long, and the ratio of aperture plus canal to total length is approximately 0.5. A single subsutural cord of a continuous reddish-brown color is present, followed anteriorly by a white area with flattened spiral threads. In occasional specimens, one of the threads is light brown in color. The large peripheral, highly gemmate sinus cord is white in the early spire whorls, with the interspaces on the peripheral cord between gemmules becoming brown colored in later whorls. On the body whorl, these interspaces become brown to reddish-brown, with raised white gemmules. Body whorl has about six primary spiral cords in addition to the peripheral sinus cord, each with an irregular pattern of golden or reddish-brown dots and white raised dashes. Cords on the body whorl are much less prominent than the sinus cord. On the sixth spiral cord there is usually a second sinus that forms on the lip. Canal is covered by numerous white spiral cords, which are more flattened and less colored than the body whorl cords. The combination of the prominently gemmate sinus cord with white gemmae and golden or reddish-brown interspaces which are white in early spire whorls and become more deeply colored as the shell whorl gets larger. The uniformly colored subsutural cord, and the presence of a second sinus are distinctive characteristics of the species. Adult shells are 36–49 mm in length. Because of the deep habitats, little is known about the feeding habits of this species. These snails are venomous with disulfide-rich polypeptides in their venom ducts. These bioactive peptides become a resource for novel pharmacologically active compounds.

Toxin Type: Turripeptides – P-like (three disulfides): Gsg 9.1 (Kastin, 2013; Aguilar et al., 200; King, 2015).

Gemmula peraspera (Marwick, 1931)

Apertural view

Common Name: Not designated

Geographical Distribution: New Zealand

Habitat: Deep-water species

Identifying Features: This species is somewhat similar to the recently described new species viz. Gemmula disjuncta, but is taller and has sharper tubercles set on a sharper keel. Shell is tall and narrow, with a narrow spire of 0.5–0.6 total height. Whorls are prominently keeled at about mid-height on spire, forming narrow, strongly concave sutural ramp. Sculpture of peripheral row is of small nodules, quite sharply pointed on most specimens, 16–20 per whorl. Nodules are formed at apex of anal sinus, where growth lines curve sharply forward after their backward inclination over sutural ramp. Spiral sculpture is of 2–3 prominent, widely spaced cords on base. Anterior canal is long and narrow, deflected weakly to left, but incomplete on all known specimens. Protoconch is narrowly conical and is of 5 or 6 whorls which are smooth near apex but with regular axial costae lower down. Shell size varies from 27 to 33 mm in height. Because of the deep habitats, little is known about the feeding habits of this species. These snails are venomous with disulfide-rich polypeptides in their venom ducts. These bioactive peptides become a resource for novel pharmacologically active compounds.

Toxin Type: All the species of this genus are venomous and possess toxic turripeptides; venom type not reported.

Gemmula ambara **(Olivera, Hillyard and Watkins, 2008)**

Apertural view   Abapertural view

Common Name: Not designated

Geographical Distribution: Philippines

Habitat: Deep-water species

Identifying Features: Shell of this species is moderately broad fusiform with a length of 30–55 mm. Overall color of the shell is white, flushed with a distinct violet or purplish tone. Protoconch has 3–4 translucent yellowish-brown to purplish brown whorls, axially costate over the last two-protoconch whorls. Because of the deep habitats, little is known about the feeding habits of this species. These snails are venomous with disulfide-rich polypeptides in their venom ducts. These bioactive peptides become a resource for novel pharmacologically active compounds.

Toxin Type: All the species of this genus are venomous and possess toxic turripeptides; venom type not reported.

Gemmula diomedea (Powell, 1964)

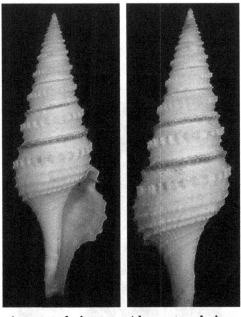

Apertural view   Abapertural view

Common Name: Albatross turrid

Geographical Distribution: Philippines; N. & W. Australia

Habitat: Deep-water species

Identifying Features: Shell size of this species varies from 40 to 90 mm in height. These snails are venomous with disulfide-rich polypeptides in their venom ducts. These bioactive peptides become a resource for novel pharmacologically active compounds. Nothing much is known about its biology.

Toxin Type: Turripeptides – P-like (three disulfides): Gdm 9.1 (Kastin, 2013; Aguilar et al., 2009; King, 2015).

Gemmula periscelida **(Dall, 1889)**

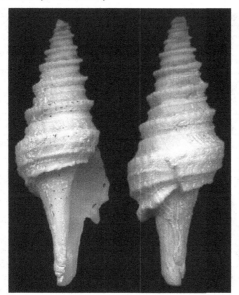

Apertural view   Abapertural view

Common Name: Atlantic Gem-turris

Geographical Distribution: S. W. of Key West, Monroe Co., Florida Keys

Habitat: From a depth of 200 fathoms

Identifying Features: Shells of this species grow to a maximum size of 45 mm. These snails are venomous with disulfide-rich polypeptides in their venom ducts. These bioactive peptides are likely to become a resource for novel pharmacologically active compounds.

Toxin Type: Turripeptide GpIAa (http://www.uniprot.org/uniprot/P0C1X3).

Gemmula mystica **(Simone, 2005)**

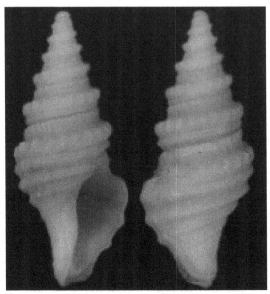

Apertural view   Abapertural view

Common Name: It is a new species and common name not designated

Geographical Distribution: Off Ubatuba, São Paulo, Brazil

Habitat: Sandy, from 510 to 530 m depth

Identifying Features: Shell of this species is turriform to fusiform. It has a spire, which is long and is about same body whorl length. Protoconch is broad, tall and is about 1/6 of body whorl width. Teleoconch is with about 6–7 convex whorls. Aperture is elliptical. Outer lip is thin. Inner lip is weakly concave and smooth. Maximum shell size is 20 mm and color is pure white. These snails are venomous with disulfide-rich polypeptides in their venom ducts. These bioactive peptides are likely to become a resource for novel pharmacologically active compounds.

Toxin Type: Turripeptides and venom type is not reported.

Gemmula lisajoni (Olivera, 1999)

Apertural view   Abapertural view

Common Name: Not designated

Geographical Distribution: Philippines, North Pacific Ocean

Habitat: Deep-water species; 50–500 m depth

Identifying Features: Maximum height of this species is 35 mm. Because of the deep habitats, little is known about the feeding habits of this species. These snails are venomous with disulfide-rich polypeptides in their venom ducts. These bioactive peptides become a resource for novel pharmacologically active compounds. Nothing much is known about its biology.

Toxin Type: Turripeptide Gli9.1 (http://www.uniprot.org/uniprot/ P0DKT5).

Lophiotoma albina (Lamarck, 1822)

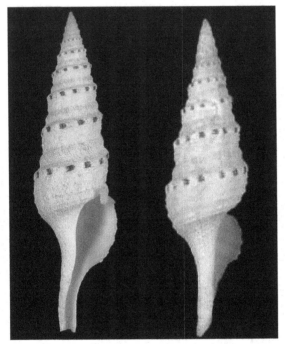

Apertural view   Abapertural view

Common Name: Not designated

Geographical Distribution: Indo-Pacific

Habitat: Deep-water habitat, 20–25 m

Identifying Features: Maximum height of this species is 75 mm. Because of the deep habitats, little is known about the feeding habits of this species. These snails are venomous with disulfide-rich polypeptides in their venom ducts. These bioactive peptides become a resource for novel pharmacologically active compounds. Nothing much is known about its biology.

Toxin Type: Turripeptide OL11-like (http://www.uniprot.org/uniprot/P0DKM9);
   Turripeptide OL55-like (http://www.uniprot.org/uniprot/P0DKP2);
   Turripeptide Lal91 (Aguilar et al., 2009);

Lophiotoma cingulifera (**Lamarck, 1822**) (= *Xenuroturris cingulifera, Iotyrris cingulifera*)

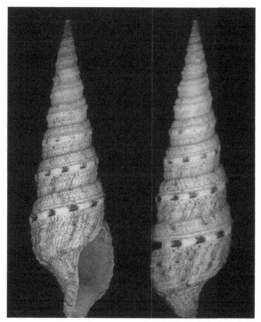

Apertural view   Abapertural view

Common Name: Necklace turrid

Geographical Distribution: Indo-West Pacific; Aldabra, Madagascar, Mascarene Basin, Mauritius, the Red Sea and Tanzania

Habitat: 50–500 m depths; coral reefs

Identifying Features: Shell of this species is fusiform and is rather narrow with a very long spire and short siphonal canal. It is corded with larger and smaller riblets and raised lines. Shell is very slightly angulated on each whorl by a somewhat larger rib, which is occasionally bipartite. Growth striae are sharp, sometimes decussating the smaller spiral lines. Color of the shell is whitish with very closely and finely peppered with chestnut and with chestnut spots on the shoulder rib. Shell size varies from 30 to 75 mm in height with a diameter of 16 mm. As these snails are venomous, they should be carefully handled.

Toxin Type: Turripeptide Lci91 (Aguilar et al., 2009).

Lophiotoma indica (Röding, 1798)

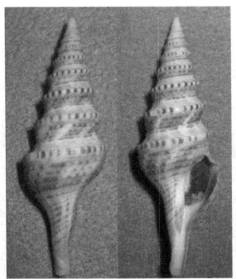

Abapertural view   Apertural view

Common Name: Indian turrid

Geographical Distribution: Indo-West Pacific: from East Africa to eastern Melanesia; north to Japan and south to Queensland

Habitat: Benthic; depth range 0–50 m

Identifying Features: The fusiform shell of this species is somewhat less ridged and striated and has a long siphonal canal. Shoulder angle is very slight and the central ridge is forming a carina. Other revolving ridges are smaller and closer than other species in this genus. Color of the shell is yellowish-brown, sometimes indistinctly marbled or variegated. Size of the shell ranges from 35 to 95 mm.

Toxin Type: Turripeptide; Crude extracts of this species exhibited caseinolytic and gerlatinolytic activities. Further, these extracts have also been reported to induce moderate level of hemolysis in chicken blood (Arumugam et al., 2013).

Lophiotoma olangoensis Olivera, 2002 (= *Xenuroturris olangoensis*)

Apertural view   **Abapertural view**

Common Name: Necklace turrid

Geographical Distribution: Seas around the Philippines and Japan

Habitat: Not reported

Identifying Features: As these snails are venomous, they should be carefully handled. No other facts are available for this species

Toxin Type: Turripeptides OL105, OL108, OL11, OL127, OL135, OL139, OL142, OL172, OL184, OL22, OL25, OL38, OL47, OL49, OL55, OL57, O67L, OL71, and OL78 (King, 2015);

Turripeptides – P-like (three disulfides): OL08, OL142, OL11 and OL 135;

Turripeptides – O-like (three disulfides): Ol38 and OL25;

Turripeptides – Four disulfides: OL127, OL 47, OL105 and OL 67 (Kastin, 2013);

Turripeptide Lol9.1 (OL 11) (http://www.uniprot.org/uniprot/P0DKM7);

Turripeptide OL139 (http://www.uniprot.org/uniprot/P0DKN8);

Recombinant Toxin Lol142 (MyBiosource Inc.-http://www. MyBioSource.com).

Polystira albida (G. Perry, 1811)

Apertural view   Abapertural view

Common Name: White Giant-turris

Geographical Distribution: USA: Florida: East Florida, West Florida, Florida Keys; USA: Louisiana, Texas; Mexico: Campeche Bank; Panama, Colombia, Venezuela: Sucre, Isla Margarita; Cuba: North Matanzas; Jamaica, Puerto Rico, French Guiana, Surinam, Brazil: Amapa

Habitat: Depths of 15–79 m

Identifying Features: Shells are pure-white in color with 5 to 7 spiral smooth cords of unequal size between the well-impressed sutures. Sculpture is more distinct and smooth. Largest squarish cord is seen behind the slot-like, deep sinus. Maximum reported size (length) of the shell of this species is 116 mm. As these snails are venomous, they should be carefully handled.

Toxin Type: Turripeptides – P-like (three disulfides): pal 9a (Kastin, 2013; Aguilar et al., 2009);

Turripeptide Pal9.2 (http://www.uniprot.org/uniprot/P0DKT1);

Turripeptide PaIAa (http://www.uniprot.org/uniprot/P0C1X4).

Turris assyria **(Olivera, Seronay and Fedosov, 2010)**
Image not available.

Common Name: Not designated

Geographical Distribution: Philippines; from the Solomon Islands and Queensland, north to Japan

Habitat: Not reported

Identifying Features: Shell of this species is turriform, consisting of 12 teleoconch whorls and protoconch is missing in the holotype. Shell surface is glossy and axial sculpture is absent. First five whorls are somewhat corroded. Succeeding whorls possess four distinct spiral cords, with brown maculations, among which first and third are stronger and wider than the others. Third spiral cord is notably thickened, forming a distinct keel that marks the periphery and gives the whorl its characteristic angular shape. Last whorl and canal bear 12 major maculated spiral cords, which are separated by minor sharp intermediate white threads, with additional white threads towards the tip of the canal. Siphonal canal is relatively long and narrow, and the distinct cords with maculations in the canal are darker brown in color than in the rest of the body whorl. As these snails are venomous, they should be carefully handled.

Toxin Type: Turripeptides; venom type not reported.

Turris babylonia (Linnaeus, 1758)

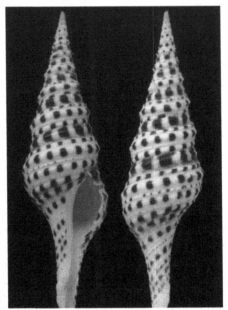

Apertural view   **Abapertural view**

Common Name: Babylon turrid

Geographical Distribution: Pacific Ocean along the Philippines, Indonesia, the Solomon Islands, Papua New Guinea, Timor; in the Indian Ocean off Mauritius and the Mascarene Basin

Habitat: Soft substrates

Identifying Features: In this species, coloration in foot is darker gray with small irregular darker spots. Region from the sinus cord to the subsutural cord is brown to off white. This species has also a very thin sinus cord and an extremely broad and boldly maculated subsutural cord, which is with very much bolder markings. Individual whorls are generally less bulbous and this species grows up to 10 cm.

Toxin Type: Turripeptides; venom type not reported.

3.3 AUGER (TEREBRID) SNAILS (FAMILY: TEREBRIDAE)

This family includes auger snails (auger shells). Typically, shells of this family are shaped like long, slender augers or screws. Aperture of the shells is irregular with a short anterior canal or notch. Further, shells are flattened and more twisted than spiraled. One or two folds are seen on the columella. World-wide, there are about 300 species of family Terebridae. Species in this family are predominantly grouped in either the Terebra or the Hastula genus, with a few remaining in two other genera. All are sand-dwelling carnivores found in warmer waters. By projecting a venomous barb like that of the cone shell molluscs, they stun their prey, which consists of various marine worms (http://shells.tricity.wsu.edu/ArcherdShellCollection/Gastropoda/Terebridae.html).

For members of the family terebrida characterized by elongated shell with lots of turns. Sculpture is represented by the outer surface of the shells spiral ribs, grooves. The structure of the family terebrida includes about 265-year-species, mostly common in the Indo-Pacific. In addition, the district representatives of this family inhabit tropical sea eastern Pacific and Atlantic Oceans, as well as moderate sea off the coast of Australia, New Zealand, North and South America. Terebridy inhabit sandy and muddy bottoms, coral reefs for small and medium depths. Only a few species have adapted to life at depths of about 350 m. Terebridy – predators. Its prey – a variety of worms they find in the sand. Some terebrid has poison gland to immobilize prey.

Hastula hectica **(Linnaeus, 1758)**

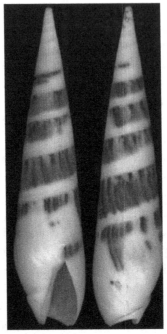

Apertural view   Abapertural view

Common Name: Sandbeach auger, hectic auger

Geographical Distribution: Hawaii and Indo-Pacific

Habitat: A rare species found along beach slopes below the high water mark

Identifying Features: Spire is often damaged and extent of dark banding is variable. It feeds on worms at night using its broad foot as a sail to surf up and down the sandy slope with successive waves, quickly burying as the water recedes. It has a striking radular tooth to inject venom that looks like a perforated spear. Length of the shell varies between 30 mm and 80 mm. As these snails are venomous, they should be carefully handled.

Toxin Type: Venom of this species contains a large complement of small disulfide-rich peptides (Imperial et al., 2007);

Augerpeptides hhe1a, Hhe53, Hhe6.1, Hhe6.2, Hhe6.3, Hhe6.4, hhe7a, Hhe9, Hhe9.2, hhe9a, hheTx1, hheTx2, hheTx3, hheTx4 and hheTx5 (King, 2015);

Augerpeptides – Two disulfides: hhe Tx1, hhe 1a, Hhe 9 and Hhe 53;

Augerpeptides – P-like (three disulfides): hhe 9a and Hhe 9.2;

Augerpeptides – O-like (three disulfides): hhe7a, Hhe 6.2, Hhe 6.3, Hhe 6.4 and Hhe 6.1;

Augerpeptides – Four disulfides: hhe Tx2, hhe Tx3, hhe Tx3, hhe Tx4 and hhe Tx5 (Kastin, 2013);

Augerpeptide – hhe9a, Hhe92 (Aguilar et al., 2009);

Conotoxin-like peptides-aug toxins, Agx-s7a, Agx-s11a; Hhe1a, HheTx2, HheTx1 C Hhe7a, Hhe9a, HheTx3, HheTx4, HheTx5;

Augerpeptide Hhe53.

Terebra consobrina (Deshayes, 1857)

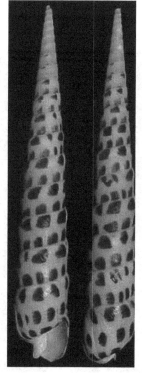

Apertural view   Abapertural view

Common Name: Not designated

Geographical Distribution: Red Sea – Madagascar; Indian Ocean

Habitat: Not reported

Identifying Features: Color of the shell is cream, which is ornamented with 2 rows of squarish brown spots, 3 on body whorl. Outline of whorls is straight and is very slightly turreted. Aperture is quadrate and columella is short and recurved. Shell size varies form 70 to 135 mm. As these snails are venomous, they should be carefully handled. Venom glands and harpoon-like radula teeth are present in this species.

Toxin Type: Pharmacological study on extracts of venom glands of this species revealed high inhibitory activities to nAChRs predominantly of the neuronal subtype. The observed effects are similar to those produced by α-conopeptides from the venom of cone snails (Kendel et al., 2013); Peptides of the A-superfamily (Kendel et al., 2013).

Terebra argus **(Hinds, 1844)**

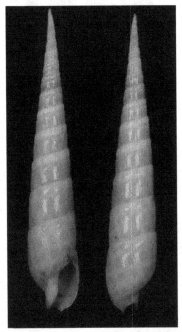

Apertural view   Abapertural view

Common Name: Argus Auger

Geographical Distribution: Red Sea – E. Africa – Hawaii

Habitat: In slightly muddy and silty sand shallow waters

Identifying Features: Color of the shell is ivory. Shell is usually with 3 rows of faint nebulous yellow spots per whorl, 4 on body whorl. Outline of whorls are straight and protoconch is with three and half whorls, the last one being unusually long. Aperture is quadrzte and columella is recurved. Shell size varies from 45 to 113 mm. This species possesses venom glands and harpoon-like radula teeth. As these snails are venomous, they should be carefully handled.

Toxin Type: Pharmacological study on extracts of venom glands of this species revealed high inhibitory activities to nAChRs predominantly of the neuronal subtype. The observed effects are similar to those produced by α-conopeptides from the venom of cone snails (Kendel et al., 2013); Peptides of the A-superfamily (Kendel et al., 2013).

Terebra variegata **(Gray, 1834)** (= *Terebra (Variegata-group) variegata*)

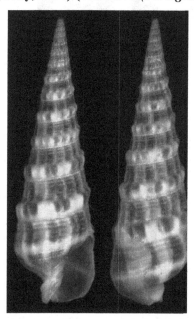

Apertural view   **Abapertural view**

Common Name: Variegated auger

Geographical Distribution: Pacific Ocean from Baja California Peninsula to Peru; Off Galápagos Islands

Habitat: Intertidal and to a depth of 110 m

Identifying Features: Shell of this species typically is flat-sided, grayish with a brown- and white-spotted subsututal band and a white peripheral band that shows through the aperture as a stripe. Streaks and spots of color mark the surface. Aperture is quadrate. Columella is recurved with two implications. Length of the shell varies between 25 mm and 100 mm and its dia. is 19.3 mm. As these snails are venomous, they should be carefully handled.

Toxin Type: Peptide Tv1, which has an M-lek conopeptide arrangement (King, 2015).

Terebra subulata **(Linnaeus, C., 1758)**

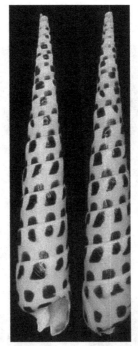

Apertural view   Abapertural view

Common Name: Chocolate spotted auger, Subulate auger

Geographical Distribution: Pacific and Indian oceans; New Guinea; Solomons; Fiji; Indo-W Pacific

Habitat: Tropical corner to local shallow sea of sand or sand-and-mud bottom

Identifying Features: Shoulder turns of this species is well pronounced and convex. Siphon outgrowth is relatively short and slightly curved, with a wide opening. Outer lip is not thickened and mouth is oval. Spiral sculpture is weakly expressed with much smoother edges. Shell color is yellowish or white. Surface of the shell is covered with brown blots of rectangular shape extending in a spiral. Mouth of a yellowish or white. Shell size varies from 70 to 200 mm. This is a carnivorous species. As these snails are venomous, they should be carefully handled.

Toxin Type: Augerpeptides – O-like (three disulfides): Agx 56a and Age 57a; Augerpeptides – Four disulfides: Agx s11a (Kastin, 2013); Augertoxins s6a, s7a, s11 and s11a (Imperial, 2007, King, 2015; www.hoelzel-biotech.com, 2014).

KEYWORDS

- **auger (terebrid) snails**
- **cone snails**
- **conotoxin**
- **turrid snails**
- **turripeptide**
- **ziconotide**

CONOTOXINS

CONTENTS

4.1 INTRODUCTION

All the 600 living species of cone snails have a highly sophisticated venom production apparatus and delivery system. The venom gland that runs along the foregut synthesizes and secretes the conotoxins (conopeptides). These toxins are small, 15–40 amino acids long, and consist of a conserved core pattern with a highly variable region that is responsible for the multitude of effects conotoxins have. All such toxins are grouped together to form a venom. Most of the conotoxins are the usual paralyzing neurotoxins, but there are some with wider effects, such as sleep inducers. The *conus* species use their venoms for multiple purposes, including prey capture and defense. Each of these species can produce 50–200 different conotoxins (or 'conopeptides') and all such toxins are species-specific (Terlau & Olivera, 2004). Coinciding with their characterization, synthesis and bioengineering, these peptides have operated as pharmacological probes to dissect ion channel and receptor functions since the 1970s and have now transitioned into analgesic pharmaceuticals (http://www.dehs.umn.edu/PDFs/Conotoxins.pdf). However, in contrast to the impressive progress made in understanding the toxicology and ecology of cone snails, very few studies have been performed on the venoms of Terebridae and Turridae.

4.2 CHARACTERISTICS OF CONOTOXINS

4.2.1 VENOM APPEARANCE

Dissected venom gland extracts can be opaque, milky white to sulfur yellow in color. The milked venom is clear, unless hydrophobic peptides, such as the δ-conotoxins, are present. Milked venom consists to a lesser extent of proteins and low molecular mass organic compounds, with peptides being the dominant constituent; the majority of the milked venom volume is equivalent to seawater.

4.2.2 VENOM SOLUBILITY

Synthetic and extracted conotoxins/conopeptides are soluble in water, producing a slightly translucent solution that may foam if agitated. Native venoms contain small insoluble particles or granules (http://www.dehs.umn.edu/PDFs/Conotoxins.pdf).

4.2.3 VENOM STABILITY

When stored correctly under laboratory conditions, conotoxins/conopeptides are highly stable identities. Due to disulfide bonding, a high level of intrinsic structural stability is present within the conotoxins. Yet, as peptides, they are not resistant to enzymatic digestions, microbial breakdown, disulfide reduction and/or chemical oxidation. Any chemical modification to the native peptide(s) will typically lead to a decrease or removal of biological activity. Although heating leads to their degradation, prolonged exposure to heat, such as autoclaving alone, is regarded as an ineffective measure to completely remove biological functionality (http://www.dehs.umn.edu/PDFs/Conotoxins.pdf).

4.2.4 DEFINITION OF CONOTOXINS

The conotoxins are paralytic poisons from cone snails that block the trans mission of a nerve impulse from the nerve to the muscle at the neuro muscular junction.

4.2.5 CLASSIFICATION OF CONOTOXINS

Classification of these toxins has been done by different workers in different ways. According to Terlau and Olivera (2004) conopeptides of cone snails fall within 15 superfamilies with stereotypical structures. Lewis et al. (2012) classified these toxins into 16 superfamilies along with the families, number of species and feeding habits of each superfamily as given in the following table.

Superfamily	Family	No. of species	Feeding habit
A	α, ρ, k	43	VMP
D	α	6	V
I1	i	11	VMP
I2	k	17	VMP
I3	–	3	V
J	k	2	V
L	α	2	VP
M	α, i, k, μ	28	VMP
O1	δ, k, μO, ω	53	VMP
O2	γ	15	VMP
O3	–	6	VMP
P	–	6	VM
S	α, σ	6	VMP
T	×, μ, Σ	24	VMP
V	–	2	V
Y	–	1	V

P, fish; M, molluscs; V, worms.

Source: Lewis et al. (2012).

Bingham et al. (http://www.dehs.umn.edu/PDFs/Conotoxins.pdf) classified the conotoxins into 7 superfamilies' along with their families as given in the following table.

Superfamily	Families of conotoxins
A-Superfamily	α-, αA-, κA- and ρ
M-Superfamily	ψ-, μ- and κM
O-Superfamily	μO-, δ-, ω-, κ-, and γ
P-Superfamily	Unknown
S-Superfamily	Unknown
T-Superfamily	τ- and χ
I-Superfamily	Unknown

Source: Bingham et al., http://www.dehs.umn.edu/PDFs/Conotoxins.pdf.

Samundeswari et al. (2014) classified the conotoxins into α, δ, γ, δ, κ, ω, ρ, χ, σ, μ, conopressin, conantokin, contulakin.

Anon (http://www.avru.org/health/health_cones.html) classified the conotoxins into 10 families along with their mode of action as given in the following table.

Family	Mode of Action
a-conotoxins	Competitively block muscle and vertebrate neuronal nicotinic ACh receptors
g-conotoxins	Activate pacemaker cationic channel
d-conotoxins	Activate predominantly mollusc sodium channels
k-conotoxins	Block potassium channels
m-conotoxins	Block vertebrate muscle/nerve sodium channels
s-conotoxins	Inhibit 5HT3 channel
y-conotoxins	Non-competitively block muscle ACh receptors
w-conotoxins	Block N-type or P/Q-type vertebrate calcium channels
Conopressin	Vasopressin agonist
Conantokin	Inhibit vertebrate NMDA-glutamate channels

http://www.avru.org/health/health_cones.html.

Anderson and Bokor (2012) reported on the physiological effects of the different families of conotoxins.

Family	Physiological Effects
Alpha (α)	Blocks nicotinic receptors. Produces muscle paralysis.
delta (δ)	Inhibits the fast inactivation of voltage gated sodium channels.
epsilon (ε)	Affects presynaptic calcium channels needed for action potential activity.
iota (ι)	Agonist at sodium gated channels with no delayed inactivation.
kappa (κ)	Antagonist of potassium gated channels. Interferes with repolarization.
mu (μ)	Antagonist of sodium gated channels.
rho (ρ)	Impacts alpha-adrenal receptors affecting blood pressure and smooth muscle.
sigma (σ)	Affects serotonin activity. Impacts mood, appetite and stress control.
chi (χ)	Affects neuronal adrenergic transporter.
omega (ω)	Works on voltage gated calcium channels.
Conantokins	Antagonize glutamate, the main excitatory neurotransmitter in the brain, at N-methyl-D-aspartate receptors.
Conopressins	Modulate vasopressin/oxytocin receptors. Increases blood pressure.

Source: Anderson and Bokor (2012).

Olivera and Cruz (2001) reported on the pharmacological families and actions of different superfamilies of cononotoxins as given below.

Superfamily	Pharmacological family	
	Designation	Action
A	α	Competitive antagonist of ACh receptor
	α A	Competitive antagonist of ACh receptor
	kA	Inhibits K+ channels (VSPC)
M	μ	Blocks Na+ channels (VSSC) at site I
	ψ	Noncompetitive antagonist of ACh receptor
O	ω	Blocks Ca++ channels (VSCC)
	ω o	Blocks Ca++ channels (VSCC)
	k	Inhibits K+ channel

CONTINUED

Superfamily	Pharmacological family	
	Designation	Action
	δ	Delays inactivation of VSSC by binding to site VI
	μO	Blocks VSCC; does not compete with TTX and STX
P		Unknown
S	σ	5-HT3 receptor
T		Unknown

VSPC: voltage-sensitive potassium channel; VSSC: Voltage- sensitive sodium channel. VSCC: Volt-age-sensitive calcium channel; TTX: Tetrodotoxin; STX: Saxitoxin.

Source: Olivera and Cruz (2001).

Venoms from cone snails (Conidae) have been extensively studied during the last decades, but those from other members of families Terebridae and Turridae have attracted less interest so far. The venom from Terebridae (called teretoxins) are similar in structure and composition to the venom (conotoxins) from cone shells. The extracts of terebrids have been reported to exhibit remarkably high inhibitory activities on almost all nAChRs tested, in particular on the α7 subtype suggesting the presence of peptides of the A-superfamily from the venom of *Conus* species. In contrast, no effects on the potassium and sodium channels tested were observed. The venoms of terebrid snails may therefore offer an additional source of novel biologically active peptides (Kendel et al., 2013; http://txmarspecies. tamug.edu/invertfamilydetails.cfm?famnameID=Terebridae). While only a few teretoxins have been described in the literature, results from preliminary characterizations indicate their potential as biochemical tools for analyzing the mechanics and function of the neuronal circuit. Several teretoxins, previously referred to as augertoxins, identified from terebrid species viz. *Terebra subulata* and *Hastula hectica* have a cysteine framework similar to the O-superfamily of conotoxins. This suggests that they may fold into the inhibitory cysteine knot motif referred to as the ICK motif. The ICK motif is common among peptide toxins from various organisms including snakes and spiders, and is known to block ion channels. While the *T. subulata* teretoxins identified have a similar O-superfamily cysteine framework, the signal sequence of the precursor region is not homologous with the conotoxin O-superfamily signal sequence. This suggests that although the mature toxins are similar, the genes encoding the peptides are not. Likewise, the teretoxins identified from *H. hectica* have cysteine patterns similar to the O and P conotoxin superfamilies, but their signal sequences are highly divergent. These findings indicate the genetic makeup of conus and terebrid toxins are not the same. It thus follows that newly discovered teretoxins could have diverse functional applications compared to their conotoxin counterparts (Puillandre and Holford, 2010).

KEYWORDS

- conopeptides
- conotoxins
- milked venom
- pharmacological probes
- teretoxins
- venom stability

CHAPTER 5

ENVENOMATION OF CONE SNAILS

CONTENTS

5.1   VENOM ACTION OF THE CONUS SNAILS

The venom of the *Conus* species is composed of conotoxins, which are neurotoxins of low molecular weight. The geographic cone snail, which is one of the most venomous creatures on the earth and is known to have killed dozens of people in accidental encounters No antidote exists for a cone snail sting. The action of conotoxins is extremely fast, which is compatible with the slowness of the snail in its environment and the consequent difficulty involved in capturing the poisoned prey. The action of the conotoxins occurs by blockage of muscular and neural receptors. There are two different toxin effects in the venom. The first, the "lightning-strike" effect, causes immediate immobilization of the injected prey through peptides that inhibit voltage-gated sodium channel inactivation, as well as peptides that block potassium channels. Together, this combination results in a massive depolarization of any axons in the immediate vicinity of the venom injection site, causing an effect similar to electrocuting the prey. The second effect is achieved more slowly and involves total inhibition of neuromuscular transmission through conopeptides, which act at sites remote from the venom injection site, such as neuromuscular junctions.

5.2   CONOTOXIN VENOM DELIVERY

Conotoxins block ion channels within the nervous system by interrupting the chemical signals. Cones produce more than one hundred different varieties of toxins, which target specific muscle groups. Some toxins affect skeletal muscles and others affect major organs, such as the heart. Seconds after the sting, communication stops between the muscle cells, causing immediate paralysis.

5.3   CLINICAL ASPECTS OF THE ENVENOMATION

The exact number of proven deaths caused by envenomation by the Conus shells is not known, though about 50 supposed occurrences have been reported. Envenomation can simulate other causes of death, like myocardial infarction or cerebral ischemia, and may not always be correctly identified. A typical accident involving Conus shells, initially presents intense burning pain at the site of the sting, which evolves in about one hour to progressive paralysis of the body muscles without other local symptoms or signs. In later phases, the patient can develop palpebral ptosis, blurry vision and speech and deglutition difficulties, unconsciousness and dyspnea, with possible evolution to respiratory arrest that can be fatal and occurs 40 min to 5 h after the sting. Diagnosis of the injuries caused by venomous molluscs is based on clinical observation. There is no antidote for the toxins and severe envenomation should be treated with artificial ventilation, the only effective measure. Severity depends on the time passed since envenomation and of the full installation of the manifestations, with some consideration given to the sting site. A lack of medical resources also influences the prognosis.

5.4 CONE SNAIL STING SYMPTOMS

• Most stings occur on the hands and fingers due to handling.
• Mild stings are similar to a wasp or bee sting with localized burning and sharp stinging symptoms. They can be intense and also have numbness and tingling to the wounded area.
• Some sting symptoms can progress to include cyanosis (blueness at the site due to decreased blood flow), and even numbness or tingling involving an entire limb.
• Severe cases show total limb numbness that progress to the area around the mouth (perioral) and then the entire body. Paralysis (inability to move a part or entire body) can occur leading to paralysis of the diaphragm, which stops the ability to breath.
• Coma and death can in severe cases where the diaphragm is paralyzed.
• Other symptoms that can occur include: fainting (syncope), itching, loss of coordination, heart failure, difficulty speaking, difficulty breathing, and double vision.
• Symptoms can begin within minutes or take days to appear after the venom is injected.

5.5 CONE SNAIL STING TREATMENT

• If SCUBA diving, the diver stung should safely surface immediately accompanied by another diver.
• There is no antivenom available for cone snail stings.
• Pressure immobilization technique may be used.
• Elastic bandage (similar to ACE bandage) may be used to wrap the limb starting at the distal end (fingers or toes) and to wrap toward the body. It should be tight but the fingers and toes should remain pink so that the circulation is not cut off.
• The extremity should also be immobilized with a splint or stick of some sort to prevent it from bending at the joints.
• The elastic bandage should be removed for 90 seconds every 10 min and then reapplied for the first 4 to 6 h (hopefully medical care can be received within this time period).

5.6   OTHER TREATMENT OPTIONS

The other treatment options that may help include:

- Affected area may be immersed in water as hot as is tolerable (water temperature not to exceed 140°F or 60°C).
- A local anesthetic is injected into the wound area.
- Some reports suggest that Edrophonium (Enlon, Tensilon) 10 mg may be used as therapy for paralysis. A 2-mg test-dose should first be administered, and if effective, followed by an additional 8-mg dose. Atropine (Atreza, Sal-Tropine) 0.6 mg should be immediately available for intravenous administration in case of an adverse reaction to edrophonium.
- A 2 to 4 mg dose naloxone (Narcan) given may help treat severe hypotension (low blood pressure).
- Local excision (cutting out the area stung) by a health care professional (controversial and not widely recommended).
- Incision and suction (controversial and not widely recommended).
- Excessive movement may be avoided and the patient may be kept calm and warm.
- Artificial respiration may save the person's life.

5.7   CONE SNAIL STING PREVENTION

- Picking up cone shells from beach or intertidal waters should be avoided; if a person does pick up a shell, they should be wearing proper gloves and carefully grasp the large end of the shell.
- If any part of the snail begins to stick out from the shell, the cone should be dropped immediately.
- Shell should be carried by the large end of the shell.
- Shell should not be carried inside a wet suit, clothing pocket, or dive suite buoyancy compensator.
- Medical Care and treatment should be sought at the right time/as soon as possible.
- Intensive care hospitalization, including use of a respirator, may be required.

KEYWORDS

- atropine
- edrophonium
- envenomation
- Narcan
- sting symptoms
- venom delivery

CHAPTER 6

THERAPEUTIC USES OF CONE SNAIL VENOMS

CONTENTS

6.1 INTRODUCTION

Each species of cone snail has been reported to produce in excess of 1000
conopeptides, with an estimated 5% overlap in conopeptides found between
species. The small molecular size of conopeptides (typically, 5 kDa) and
their relative ease of synthesis, structural stability and target specificity
make them ideal pharmacological probes. The broadly evolved bioactivity
of conopeptides provides a unique source of new research tools and
potential therapeutic agents. These predatory animals produce an
estimated approximately 100,000 distinct conotoxins many therapeutically
interesting molecules remain to be isolated and characterized from the genus
Conus. At the moment, only 0.1% of conopeptides have been characterized
pharmacologically, yet many have already been identified with clinical
potential (Lewis et al., 2012). Over two decades of research on venom
peptides derived from cone snails ("conopeptides or conotoxins") has led
to several compounds that have reached human clinical trials, most of them
for the treatment of pain. Remarkably, none of the conopeptides in clinical
development mediate analgesia through the opioid receptors, underlying
the diverse and novel neuropharmacology evolved by Conus snails. The
conopeptides studied to-date in animal models, have exhibited antinociceptive, antiepileptic, neuroprotective or cardioprotective activities. Screening
results also suggest applications of conotoxins in cancer, neuromuscular
and psychiatric disorders. Additional potentially important applications of
conotoxin research are the discovery and validation of new therapeutic
targets, also defining novel-binding sites on already validated molecular
targets. As the structural and functional diversity of conotoxins is being
investigated, the Conus venoms continue to surprise with the plethora
of neuropharmacological compounds and potential new therapeutics.
This review summarizes recent efforts in the discovery of conopeptides,
and their preclinical and clinical development.

6.2 MODE OF CONOTOXINS ACTIONS

Voltage-Gated Ion Channels Targeted by Conotoxins
Na+ Channel Inhibitors
Ca2+ Channel Inhibitors
K+ Channel Inhibitors
Ligand-Gated Ion Channels Targeted by Conotoxins
nAChR Inhibitors
Sodium channel targeting conotoxins
(Lt5d, Lt6c, TIIIA, Cd12a, Cd12b, BuIIIA, BuIIIB, BuIIIC, SIIIAB)
Calcium channel targeting conotoxins
(CalTx, FVIA)
Potassium channel
(Sr11a, RIIIj)
nAChR targeting conotoxins
(AlphaD-cap, AlphaD-mus, a-PIB, SrIA, SrIB, Pul4a, PrIIIE, ArIA, ArIB, Ac1.1a, Ac1.1b, PrXA, αTxIA, TxIA (AIOL)

6.3 CLINICAL POTENTIALS

Essack et al. (2012) and Lewis et al. (2012) studied on the clinical potential and mode of action of different classes of conotoxins as given in Table 6.1.

TABLE 6.1 Major Classes of Conotoxins and Their Mode of Action and Clinical Potentials

Class	Mode of action	Clinical potential
ω	Ca,2.2 inhibitor	Pain (intrathecal; phase IV)
μ	Na, inhibitor	Pain (intravenous)
μO	Na,1.8 inhibitor	Pain (intrathecal/intravenous)
δ	Na, enhancer	Unknown
k	K, inhibitor	Cardiac reperfusion
×	NET inhibitor	Pain (intrathecal; phase II)
α	nAChR inhibitor	Pain (intravenous)
σ	5HT3 receptor	Unknown

TABLE 6.1   Continued

Class	Mode of action	Clinical potential
ρ	α1-Adrenoceptor inhibitor	Cardiovascular/BPH
Conantokin	NMDA-R antagonist	Pain/epilepsy (intrathecal)
Conopressin	Vasopressin-R agonist	Cardiovascular/mood
Contulakin	Neurotensin-R agonist	Pain (intrathecal)

Source: Lewis et al. (2012).

6.4   POTENTIALLY HELPFUL CHEMICALS IN CONE SNAIL VENOM

6.4.1   ZICONOTIDE FOR PAIN RELIEF

After studying a conopeptide in the venom of a cone snail known as *Conus magus*, researchers have made a synthetic version of the peptide. The artificial chemical, called ziconotide, has some very useful properties. It has been approved as a medication in the United States by the Food and Drug Administration (FDA) and is in current use as an analgesic. This drug is up to 1,000 times more effective than morphine at relieving pain and has the added advantage of not being addictive. In addition, it doesn't cause the development of tolerance in the patient. Ziconotide is sold under the brand name of Prialt. This drug, however, has some drawbacks. It must be injected into the spinal canal (an intrathecal injection) in order to work. This is generally done continuously via an infusion pump. Serious side effects aren't common, but they do occur. One possible side effect is a severe mood change, including depression. Ziconotide is used after other analgesics have been tried and have failed to work. It is prescribed only for people who are suffering from intense and prolonged pain, such as the pain that may be experienced by people with certain types of cancer or for people experiencing neuropathic pain.

6.4.2   A NEW ANALGESIC

In 2013, the discovery of a new conopeptide with pain relieving properties was announced by Australian scientists. This chemical is said to be 100 times more effective than common analgesics and unlike Ziconotide can be taken by mouth. The researchers say that it isn't addictive. Once again, the scientists have designed a synthetic drug that resembles the natural one. At the moment the new analgesic is being tested on rats. The researchers are trying to get government approval and funding for clinical tests on humans.

6.4.3   A POSSIBLE AID FOR EPILEPSY

The best known member of the family is conantokin-G from the geography (or geographic) cone snail. Conantokins are sometimes called "sleeper peptides" because when they are injected into the brain of young mice they trigger sleep. These peptides work by a mechanism that may be helpful for people with epilepsy. As is the case with other cone snail venoms, researchers have produced synthetic molecules based on the natural ones in order to improve the properties of the compounds for medical use (http://hubpages.com/hub/Cone-Snails-Dangerous-Venom-With-Medicinal-Uses).

6.4.4   CONOTOXINS FOR HUNTINGTON'S DISEASE

Huntington's disease (HD) is a genetic disorder with autosomal dominant inheritance with progressive degeneration of neurons. It is characterized by affective, cognitive, behavioral, and motor dysfunctions. Certain cono-peptides have been reported to selectively inhibit the function of ion channels and excitatory amino acid receptors such as NMDA involved in the transmission of nerve signals in animals representing an extensive array of ion channel blockers each showing a high degree of selectivity for particular types of channels. Here we hypothesis the protective effect of certain conotoxins, of *Conus geographus* namely w-conotoxin and conantoxin, against excitotoxic neuronal cell death using 3 NP induced Huntington's model (Bhosle and Vaibhav, 2013).

6.5   BETTER MEDICAL INSULINS FOR DIABETICS

The variety of insulin used by the cone snails is also a surprise, as it is the smallest known insulin molecule yet discovered, and may offer insights into the structure and function of insulin that are useful to humans. "The cone snail venom insulin probably only has one function or goal, and that is to incapacitate prey as efficiently and as rapidly as possible. The insulin in cone snail venoms appear tailored for high potency and rapid activity. Millions of diabetics around the world use insulin as a drug, and understanding these specialized venom insulin could lead to the development of new, improved and more highly tuned insulin (http://www.australiangeographic.com.au/news/2015/01/insulin-a-secret-ingredient-in-cone-snail-venom) (Table 6.2).

TABLE 6.2   Conus Venom Peptides and Their Therapeutic Potential with Special Reference to Their Molecular Target

Molecular target	Conus peptide	Therapeutic application	Development stage reached
N-type Ca channel	ω-MVIIA	Pain	Approved by FDA 12/04
(Cav2.2)	(Prialt; zicontide)		
N-type Ca channel	ω-CVID		
(Cav2.2)	(AM336)	Pain	Phase 1
Neurotensin receptor	Contulakin-G		
	(CGX-1160)	Pain	Phase 1
Norepinephrine-transporter	χ-MrIA (derivative)		
	(Xen-2174)	Pain	Phase 1
Nicotinic receptors	α-Vc1.1 (ACV-1)	Pain	Phase 1
NMDA* receptors	Conantokin-G		
	(CGX-1007)	Epilepsy; pain	Phase 1
K+ channels	κ-PVIIA		
(Kv1 subfamily)	(CGX-1051)	Myocardial infarction	Pre-clinical
Na+ channels	μO-MrVIB		
	(CGX-1002)	Pain	Preclinical

* N-Methyl-D-aspartate.

Source: Olivera (2006).

6.6 THERAPEUTIC APPLICATIONS OF CONOTOXINS WITH SPECIAL REFERENCE TO PAIN, STROKE, NEUROMUSCULAR BLOCK, AND CARDIOPROTECTION

6.6.1 PAIN

Conotoxins act at various locations in the sensory systems that mediate pain, including the periphery, the spinal levels, and higher CNS centers. One conotoxin has been approved for use as an analgesic, several other conotoxins are in development, and still more are being explored for this indication. Certain analgesic conopeptides identified and their sources are given in Table 6.3.

TABLE 6.3 Analgesic Conopeptides and Their Sources

Name	*Conus* Species
MVIIA (Prialt®)	*C. magus*
CVID (AM336)	*C. catus*
Contulakin-G (CGX-1160)	*C. geographus*
MrIA (Xen-2174)*	*C. marmoreus*
Conantokin-G (CGX-1007)	*C. geographus*
Vc1.1 (ACV-1)	*C. victoriae*
MrVIB (CGX-1002)	*C. marmoreus*

* Xen2174 is a more chemically stable analog of MrIA.

Source: Layer and McIntosh (2006).

6.6.1.1 Conotoxins Acting on Spinal Pain Targets

Conotoxins have also been reported to serve as safe and effective nonopioid intrathecal therapies for the treatment of chronic intractable pain.

6.6.1.2   MVIIA (Prialt®)

MVIIA (also called SNX-111, ziconotide, and most recently, Prialt®) was first isolated from the venom of C. *magus* and is a member of the ω-conotoxin family that inhibits presynaptic neurotransmitter release through blockade of N-type CaV channels. MVIIA has been demonstrated to attenuate nociception in a variety of animal models, including models of persistent pain, postoperative pain, chronic inflammatory pain and neuropathic pain. MVIIA is effective in morphine tolerant rats and prolonged (7 days) intrathecal infusion of MVIIA did not produce tolerance to its analgesic effects.

6.6.1.3   CVID (AM336)

While MVIIA is effective, its side effect profile will likely limit its clinical use. CVID, first isolated from C. *catus*, represents another ω-conotoxin selective for N-type CaV channels. Consistent with a role as an antinociceptive drug, CVID has been shown to inhibit potassium ion-evoked release of the pronociceptive neurotransmitter, substance P from rat spinal cord slices. Based on its preclinical profile to date, CVID, also called AM336, may have a superior side effect profile relative to MVIIA, and is in clinical development for chronic pain management.

6.6.1.4   Contulakin-G (CGX-1160)

CGX-1160 (Contulakin-G) is a synthetic 16 amino acid O-linked glycopeptide originally isolated from the venom of C. *geographus*. CGX-1160 appears to be an agonist at neurotensin receptors.

6.6.1.5   MrIA (Xen-2174)

Recently, χ-conopeptide MrIA (χ-MrIA), a 13-residue peptide originally isolated from the venom of C. *marmoreus has been* shown to inhibit the activity of norepinephrine transporter. Consistent with its mechanism of action, χ-MrIA has been shown to produce analgesia in several preclinical pain models.

6.6.1.6   Conantokins (CGX-1007)

Conantokins-G and -T have been found to be effective in several mouse models of persistent pain, chronic inflammatory and neuropathic pain.

6.6.1.7   Vc1.1 (ACV-1)

Interestingly, an α-conopeptide from *C. victoriae*, Vc1.1 (ACV-1), shows potent analgesia in a number of animal models. Like other α-conopeptides, Vc1.1 inhibits the binding of epibatidine to neuronal AChRs and inhibits nicotine-induced catecholamine release from bovine adrenal chromaffin cells. In vivo studies, intramuscular injection of Vc1.1 suppressed pain behaviors (mechanical hyperalgesia) in two rat models of neuropathic pain, the chronic constriction injury model and the partial sciatic nerve ligation model.

6.6.2   STROKE

6.6.2.1   MVIIA (Prialt®)

The synthetic ω-conotoxin MVIIA is effective in a variety of preclinical models of stroke. Intravenous administration of MVIIA has been found to be neuroprotective in a model of transient forebrain ischemia. Similarly, intravenous administration of MVIIA was neuroprotective in the rat middle cerebral artery occlusion (MCAo) model of focal cerebral ischemia and in a rabbit model of focal cerebral ischemia.

6.6.2.2   Conantokins (CGX-1007)

The most extensively studied conantokin, conantokin-G (also called CGX-1007), produced a dose-dependent and complete neuroprotection of primary cultures of rat cerebellar neurons against neuronal injury produced by hypoxia/hypoglycemia, NMDA, glutamate, or veratridine. Since conantokin-G is effective in animal models of stroke, reduces staurosporine-induced apoptotic cell death, up-regulates the expression of the antiapoptotic protein Bcl-2, and exhibits a favorable preclinical safety profile, conantokins may represent promising compounds for use in the acute treatment of stroke.

6.6.2.3   Neuromuscular Block

Neuromuscular blocking drugs inhibit the actions of acetylcholine on AChRs of the neuromuscular junction. They are used to provide muscle relaxation during surgery, to facilitate positive pressure ventilation during and after anesthesia in the intensive care unit, and to facilitate tracheal intubation. The α3/5 conotoxins (α-conopeptide), the primary paralytic toxins of fish hunting cone snails may represent new neuromuscular blocking drugs that lack the side-effects of depolarizing neuromuscular blockers like succinylcholine. Certain α3/5 conotoxins and their sources are given in Table 6.4.

TABLE 6.4   α3/5-Conotoxins and Their Sources

Name	*Conus* Species
GI	*C. geographus*
MI	*C. magus*
SI	*C. striatus*

6.7   CARDIOPROTECTION

κ-PVIIA, also called CGX-1051, is a 27 amino acid conopeptide originally isolated from the venom of *C. purpurascens* that blocks KV *"Shaker"* channels with high affinity. Based on these preclinical results, κ-PVIIA may represent a valuable adjunct to coronary artery thrombolytic therapy and percutaneous transluminal coronary angioplasty in the management of acute myocardial infarction. Importantly, κ-PVIIA is effective following acute intravenous injection just before the time of reperfusion.

KEYWORDS

- analgesics
- conantokins
- Huntington's disease
- nAChR inhibitors
- therapeutic applications
- therapeutic targets

REFERENCES

Adams, D. J., & Berecki, G. Mechanisms of conotoxin inhibition of N-type (Cav2.2) calcium channels. *Biochimica et Biophysica Acta (BBA) – Biomembranes* 2013, 1828, 1619–1628.

Aguilar, M. B., Chan de la Rosa, R. A., Falcon, A., Olivera, B. M., & Heimer de la Cotera, E. P. Purification, characterization, and comparison with P-conotoxin-like peptide playa from the venom of the turrid snail Polystira albida from the Gulf of Mexico. *Peptides* 2009, 30, 467–476.

Aguilar, M. B., Imperial, E. L. J. S., Falcon, A., Olivera, B. M., & Heimer de la Cotera. E. P. Putative-conotoxins in vermivorous cone snails: the case of Conus delessertii. *Peptides* 2005, 26, 23–27.

Aman et al. Supporting Information. www.pnas.org/cgi/content/short/1424435112, 9 pp.

Anderson, P. D., & Bokor, G. Conotoxins: Potential Weapons from the Sea. *J Bioterr Biodef.* 2012, 3, 120. doi: 10.4172/2157–2526.1000120.

Annadurai, D., Kasinathan, R., & Lyla, P. S. Acute toxicity effect of the venom of the marine snail, Conus zeylanicus. *J Environ Biol.* 2007, 28, 825–828.

Arumugam, M., Giji, S., Tamilmozhi, S., Kumar, S., & Balasubramanian, T. studies on biochemical and biological properties of tuttids venom (*Turricula javana and Lophiotoma indica*). Indian Journal of Geo-Marine Sciences 2013, 42, 800–806.

Azam, L., Dowell, C., Watkins, M., Stitzel, J. A., Olivera, B. M., & McIntosh, J. M. α-Conotoxin BuIA, a Novel Peptide from Conus bullatus, Distinguishes among Neuronal Nicotinic Acetylcholine Receptors. *The Journal of Biological Chemistry* 2005, 280, 80–87.

Balamurgan, K., Dey, A., & Sivakumar, G. Some neuropharamacological effects of the crude extract of Conus parvatus in mice. *Pak J Biol Sci.* 2007, 10, 4136–4139.

Balamurugan, K., Dey, A., Kalaichelva, V. K., & Raju, S. Isolation and Characterization of a Novel? Conotoxin from the Venom of Vermivorous Cone Snail *Conus parvatus. Asian J. Chem.* 2008, 20, 5082–5094.

Bernáldez, J., Román-González, S. A., Martinez, O., Jiménez, S., Vivas, O., Arenas, I., Corzo, G., Arreguín, R., García, D. E., Possani, L. D., & Xei Licea, A. A *Conus regularis* Conotoxin with a Novel Eight-Cysteine Framework Inhibits Cav2.2 Channels and Displays an Anti-Nociceptive Activity. *Mar. Drugs* 2013, 11, 1188–1202.

Bhosle, P., & Vaibhav, S., Conotoxins: Possible Therapeutic Measure for Huntington's Disease. *J. Neurol. Disord.* 2013, 1, 129. doi: 10.4172/2329–6895.1000129.

Bhuiyan, M., Anand, P., Grigoryan, A., Zhuang, J., Holford, M., & Poget, S. F. NMR Structure of VarI, a Novel Bioactive Peptide Toxin from Terebrid Marine Snails. Platform-Protein Structure. 3086-Plat, http://www.academia.edu/5870573/NMR_Structure_of_VarI_a_Novel_Bioactive_Peptide_Toxin_from_Terebrid_Marine_Snails.

Bingham, J., Likeman, R. K., Hawley, J. S., Yu, P. Y.C., & Halford, Z. A. Conotoxins. http://www.dehs.umn.edu/PDFs/Conotoxins.pdf

Bingham, J., Mitsunaga, E., & Zach, L. Bergeron drugs from slugs—Past, present and future perspectives of ω-conotoxin research. *Chemico Biological Interactionsi* 2010, 183, 1–18.

Cabang, A. B., Imperial, J. S., Gajewiak, J., Watkins, M., Corneli, P. S., Olivera, B. M., & Concepcion. G. P. Characterization of a Venom Peptide from a Crassispirid Gastropod. *Toxicon.* 2011, 58, 672–680.

Chen, P., Garrett, J., & Watkins, M. Purification and characterization of a novel excitatory peptide from Conus distans venom that defines a novel gene superfamily of conotoxins. *Taxon* 2008, 52, 139–145.

Chen, W. H., Han, Y. H., Wang, Q., Miao, X. W., Ou, L., & Shao, X. X. cDNA cloning of two novel T-superfamily conotoxins from Conus leopardus. *Acta Biochim Biophys Sin (Shanghai)* 2006, 38, 287–291.

Czerwiec, E., Kalume, D. E., Roepstorff, P., Hamb, B., Furie, B., Furie, B. C., & Stenflo, J. Novel gamma-carboxyglutamic acid-containing peptides from the venom of *Conus textile. FEBS J.* 2006, 273, 2779–2788.

Dovel, S. Discovery and structural characterization of conotoxins from the venom of vermivorous cone snails. PhD Dissertation, Florida Atlantic University Boca Raton, FL, 2010.

Duda, T. F., & Remigio, E. A. Variation and evolution of toxin gene expression patterns of six closely related venomous marine snails. *Molecular Ecology* 2008, 17, 3018–3032.

Dy, C. Y., Buczek, P., Imperial, J. S., Bulaj, G., & Horvath, M. P. Structure of conkunitzin-S1, a neurotoxin and Kunitz-fold disulfide variant from cone snail. *Acta Cryst.* 2006, 62, 980–990.

Elliger, C. A., Richmond, T. A., Lebaric, Z. N., Pierce, N. T., Sweedler, J. V., & Gilly, W. F. Diversity of conotoxin types from *Conus californicus* reflects a diversity of prey types and a novel evolutionary history. *Toxico.* 2011, 57, 311–322.

Essack, M., Bajic, V. B., & Archer, J. A.C. Conotoxins that Confer Therapeutic Possibilities. *Mar. Drugs* 2012, 10, 1244–1265.

Franco, A., Kompella, S.N; Akondi, K.,, Melaun, C., Daly, N. L., Luetje, C. W., Alewood, P. F., Craik, D. J., Adams, D. J., & Marí, F. RegIIA: an α4/7-conotoxin from the venom of *Conus regius* that potently blocks α3β4 nAChRs. *Biochemical Pharmacology* 2012, 83, 419–426.

Gonzales, D. T.T., & Saloma, C. P. A bioinformatics survey for conotoxin-like sequences in three turrid snail venom duct transcriptomes. *Toxicon* 2014, 92, 66–74.

Gowd, K. H., Sabareesh, V., Sudarslal, S., Iengar, P., Franklin, B., Fernandao, A., Dewan, K., Ramaswami, M., Sarma, S. P., Sikdar, S., Balaram, P., & Krishnan, K. S. Novel Peptides of Therapeutic Promise from Indian Conidae. *Ann. N. Y. Acad. Sci.* 2005, 1056, 462–473.

Grosso, C., Patrícia Valentão, P., Ferreres, F., & Andrade, P. B. Bioactive Marine Drugs and Marine Biomaterials for Brain Diseases. *Mar Drugs.* 2014, 12, 2539–2589.

Haegen,A.V.D.,Peigneur,S.,Dyubankova,N.,Möller,C.,Marí,F.,Diego-García,E.,Naudé,R., Lescrinier, E., Herdewijn, P., & Tytgat, J. Pc16a, the first characterized peptide from *Conus pictus* venom shows a novel disulfide connectivity. *Peptides* 2012, 34, 106–113.

Helena S., Bulaj, G., Olivera, B. M., Williamson, N. A., & Purcell, W. P. Identification of Conus Peptidylprolyl Cis-Trans Isomerases (PPIases) and Assessment of Their Role in the Oxidative Folding of Conotoxins. *The Journal of Biological Chemistry,* 2010, 285, 12735–12746.

Hill, J. M., Atkins, A. R., Loughnan, M. L., Jones, A., Adams, D. A., Martin, R. C., Richard J. Lewis, R. J., Craik, D. J., & Alewood, P. F. Conotoxin TVIIA, a novel peptide from the venom of *Conus tulipa* 1. Isolation, characterization and chemical synthesis. *Eur. J. Biochem.* 2000, 267, 4642–4648.

Imperial, J. S., Novel peptides from Conus planorbis, *Terebra subulata* and *Hastula hectica*. PhD Dissertation, University of Queensland, 2007.

Janeena, A., Lyla, P. S., & Rajendran, N. Comparative Characterization of Toxins Isolated from *Conus lorois* Kiener, 1845 and *Conus amadis* Gmelin, *RJPBCS,* 1971, 6, 1759–65, 2015.

Jiang, H., Xu, C., Wang, C., Fan, C., Zhao, T., Chen, J., & Chi, C. Two novel O-superfamily conotoxins from *Conus vexillum. Toxicon* 2006, 47, 425–436.

Jiménez-Tenorio, M. Cone radular anatomy as a proxy for phylogeny and for conotoxin diversity. http://www.researchgate.net/profile/Manuel_Jimenez-Tenorio/publication/257272448_Cone_radular_anatomy_as_a_proxy_for_phylogeny_and_for_conotoxin_diversity/links/0c960524c122926ede000000.pdf

Jimenez, E. C., Olivera, B. M., & Teiche, R. W. αC-Conotoxin PrXA: A New Family of Nicotinic Acetylcholine Receptor Antagonists. *Biochemistry* 2007, 46, 8717–8724.

Jimenez, E. C., Shetty, R. P., Lirazan, M., Walker, J. R.C., Abogadie, F. C., Yoshikami, D., Cruz, L. J., & Olivera, B. M. Novel excitatory Conus peptides define a new conotoxin superfamily. *Journal of Neurochemistry*, 2003, 85, 610–621.

Junior, V. H., Neto, J. B.P., & Cobo, V. J. Venomous mollusks: the risks of human accidents by Conus snails (Gastropoda: Conidae) in Brazil. *Revista da Sociedade Brasileira de Medicina Tropical* 2006, 39, 498–500.

Kastin, A. *Handbook of Biologically Active Peptides*. Academic Press, 2013.

Kauferstein, S., Porth, C., Kendel, Y., Wunder, C., Nicke, A., Kordis, D., Favreau, P., Koua, D., Stöcklin, R., & Mebs, D. Venomic study on cone snails (*Conus* spp.) from South Africa. *Toxicon* 2011, 57, 28–34.

Kendel, Y., Melaun, C., Kurz, A., Nicke, A., Peigneur, S., Tytgat, J., Wunder, C., Dietrich Mebs, D., & Kauferstein, S. Venomous Secretions from Marine Snails of the Terebridae Family Target Acetylcholine Receptors. *Toxins* 2013, 5, 1043–1050.

King, G. *Venoms to Drugs: Venom as a Source for the Development of Human Therapeutics*. Royal Society of Chemistry, 2015.

Kobayashi, J., Nakamura, H., Hirata, Y., & Ohizumi, Y. Isolation of eburnetoxin, a vasoactive substance from the *Conus eburneus* venom. *Life Sci.* 1982, 31, 1085–1091.

Kobayashi, J., Nakamura, H., Hirata, Y., & Ohizumi, Y. Tessulatoxin, the vasoactive protein from the venom of the marine snail *Conus tessulatus. Comp Biochem Physiol B.* 1983, 74, 381–384.

Kumar, P., Venkateshvaran, K., Srivastava, P. P., Nayak, S. K., Shivaprakash, S. M., & Chakraborty, S. K. Pharmacological studies on the venom of the marine snail *Conus lentiginosus* Reeve, 1844. *International Journal of Fisheries and Aquatic Studies* 2014, 1, 79–85.

Layer, R. T., & McIntosh, J. M. Conotoxins: Therapeutic Potential and Application. *Mar. Drugs* 2006, 4, 119–142.

Lebbe, E. K., Peigneur, S., Maiti, M., Devi, P., Ravichandran, S., Lescrinier, E., Ulens, C., Waelkens, E., D'Souza, L., Herdewijn, P., & Tytgat, J. Structure-function elucidation of a new α-conotoxin, Lo1a, from *Conus longurionis. J Biol Chem.* 2014, 289, 9573–9583.

Lebbe, E. K., Peigneur, S., Maiti, M., Mille, B. G., Devi, P., Ravichandran, S., Lescrinier, E., Waelkens, E., D'Souza, L., Herdewijn, P., & Tytgat, J. Discovery of a new subclass of α-conotoxins in the venom of *Conus australis*. *Toxicon* 2014, 91, 145–154.

Lee, S., Kim, Y., Back, S. K., Choi, H., Lee, J. Y., Jung, H. H., Ryu, J. H., Suh, H., Na, H. S., Kim, H. J., Rhim, H., & Kim, J. Analgesic effect of highly reversible ω-conotoxin FVIA on N-type Ca2+ channels. *Molecular Pain* 2010, 6, 97, 12p.

Lewis, R. J., Dutertre, S., Vetter, I., & Christie, M. J. Conus Venom Peptide Pharmacology. *Pharmacol. Rev.* 2012, 64, 259–298.

Lewis, R. J., Nielsen, K. J., Craik, D. J., Loughna, M. L., Adams, D. A., Sharpe, I. A., Tudor Luchian, T. T., Adams, D. J., Bond, T., Thomas, L., Jones, A., Matheson, J., Drinkwater, R., Andrews, P. R., & Alewood, P. F. Novel ω-Conotoxins from Conus catus Discriminate among Neuronal Calcium Channel Subtypes. *The Journal of Biological Chemistry* 2000, 275, 35335–35344.

Liu, L., Chew, G., Hawrot, E., Chi, C., & Wang, C. Two potent alpha3/5 conotoxins from piscivorous *Conus achatinus*. *Acta Biochim Biophys Sin* (Shanghai) 2007, 39, 438–444.

Liu, Z., Li, H., Liu, N., Wu, C., Jiang, J., Yue, J., Jing, Y., & Dai, Q. Diversity and evolution of conotoxins in *Conus virgo, Conus eburneus, Conus imperialis* and *Conus marmoreus* from the South China Sea. *Toxicon* 2012, 60, 982–989.

Lluisma, A. O., López-Vera, E., Bulaj, G., Watkins, M., & Olivera, B. M. Characterization of a novel ψ-conotoxin from *Conus parius* Reeve. *Toxicon* 2008, 51, 174–180.

Lluisma, A. O., Milash, B. A., Moored, B., Olivera, B. M., & Bandyopadhyaya, P. K. Novel venom peptides from the cone snail *Conus pulicarius* discovered through next-generation sequencing of its venom duct transcriptome. University of Utah Institutional Repository, 24p. http://content.lib.utah.edu/utils/getfile/collection/uspace/id/2546/filename/image.

Lopez-Veraa, E., Coteraa, E. P.H., Mailloa, M., Riesgo-Escovarb, J. R., Olivera, B. M., & Aguilar, M. B. A novel structural class of toxins: the methionine-rich peptides from the venoms of turrid marine snails (Mollusca, Conoidea). *Toxicon* 2004, 43, 365–374.

Loughnan, M. L., Nicke, A., Lawrence, N., & Lewis, R. J. Novel αD-Conopeptides and Their Precursors Identified by cDNA Cloning Define the D-Conotoxin Superfamily. *Biochemistry* 2009, 48, 3717–3729.

Luo S., Zhangsun, D., Feng, J., Wu, Y., Zhu, X., & Hu, Y. Diversity of the O-superfamily conotoxins from *Conus miles*. *J Pept Sci*. 2007, 13, 44–53.

Marshall, J., Kelley, W. P., Rubakhin, S. S., Bingham, J. P., Sweedler, J. V., & Gilly, W. F. Anatomical correlates of venom production in *Conus californicus*. *Biol Bull*. 2002, 203, 27–41.

McIntosh, J. M., Hasson, A., Spira, M. E., Gray, W. R., Li, W., Marsh, M., Hillyard, D. R., & Olivera, M. M. A New Family of Conotoxins That Blocks Voltage-gated Sodium Channels. *The Journal of Biological Chemistry* 1995, 270, 16796–16802.

McIntosh, J. M., Plazas, P. V., Watkins, M., Gomez-Casati, M. E., Olivera, B. M., & Elgoyhen, A. B. A Novel α-Conotoxin, PeIA, Cloned from *Conus pergrandis*, Discriminates between Rat α9α10 and α7 Nicotinic Cholinergic Receptors. *J. Biol. Chem.* 2005, 280, 30107–30112.

Miloslavina, A., Ebertb, C., 1 Tietzec, D., Ohlenschlägerd, O., Englertb, C., Görlachd, M., & Imhof, D. An unusual peptide from *Conus villepinii*: Synthesis, solution structure, and cardioactivity. *Peptides* 2010, 31, 1292–1300.

Möller, C., & Marí, F. A 9.3 kDa Components of the Injected Venom of *Conus purpurascens* Define a New 5-disulfide Conotoxin Framework. *Biopolymers* 2011, 96, 158–165.

Moller, F. C., Rahmankhah, S., Lauer-Fields, J., Bubis, J., Fields, G. B., & Marí, F. A Novel Conotoxin Framework with a Helix−Loop−Helix (Cs α/α) *Biochemistry* 2005, 44, 15986–15996.

Montal, M. Molecular interaction of δ-conotoxins with voltage-gated sodium channels. *FERS Letter* 2005, 579, 3881–3884.

Morales-González, D., Flores-Martínez, E. F., Zamora-Bustillos, R., Rivera-Reyes, R., Michel-Morfín, J. E., Landa-Jaime, V., Falcón, A., & Aguilar, M. Diversity of A-conotoxins of three worm-hunting cone snails (*Conus brunneus, Conus nux*, and *Conus princeps*) from the Mexican Pacific coast. *Peptides.* 2015, 68, 25–32.

Neves, J., Campos, A., Osório, H., & Antunes, A., Vitor Vasconcelos, V. Conopeptides from Cape Verde *Conus crotchii. Mar Drugs* 2013, 11, 2203–2215.

Olivera, B. M. Conus Peptides: Biodiversity-based Discovery and Exogenomics. *The Journal of Biological Chemistry*, 2006, 281, 31173–31177.

Olivera, B. M., & Cruz, L. J. Conotoxins, in retrospect. *Toxicon.* 2001, 39, 7–14.

Pak, A., Isolation and Characterization of conotoxins from the venom of *Conus planorbis* and *Conus ferrugineus.* PhD Dissertation, Florida Atlantic University Boca Raton, Florida, 2014.

Peng, C., Tang, S., Pi, C., Liu, J., Wang, F., Wang, L., Zhou, W., & Liu, A. Discovery of a novel class of conotoxin from *Conus litteratus*, lt14a, with a unique cysteine pattern. *Peptides* 2006, 27, 2174–2181.

Peters, H., O'Leary, B. C., Hawkins, J. P., Carpenter, K. E., & Roberts, C. M. Conus: First Comprehensive Conservation Red List Assessment of a Marine Gastropod Mollusc Genus. *PLoS ONE*, 2013, 8, e83353. doi: 10.1371/journal.pone.0083353; http://journals.plos.org/plosone/article?id=10.1371/journal.pone.0083353.

Puillandre, N., & Holford, M. The Terebridae and teretoxins: Combining phylogeny and anatomy for concerted discovery of bioactive compounds. *BMC Chemical Biology* 2010, 10:7. doi: 10.1186/1472-6769-10-7.

Rajesh, R. P. Novel M-Superfamily and T-Superfamily conotoxins and contryphans from the vermivorous snail *Conus figulinus. Journal of Peptide Science* 2015, 21, 29–39.

Raju, S. A novel omega conotoxin from the venom of *Conus musicus*: Isolation, characterization and chemical synthesis. *Cameroon Journal of Experimental Biology* 2007, 3, 61–69.

Ramasamy M. S., & Manikandan, S. New Pharmacological Targets From Indian Cone Snails. *Mini- Reviews in Medicinal Chemistry*, 2011, 11, 125–130.

Ramlakhan, R. E. Isolation and characterization of novel conopeptides. MS Dissertation, Florida Atlantic University, 2002.

Sabareesh, V., Gowd, K. H., Ramasamy, P., Sudarsla, S., Krishna, K. S., Sikdar, S. K., & Balaram, P. Characterization of contryphans from Conus loroisii and Conus amadis that target calcium channels. *Peptides* 2006, 27, 2647–2654.

Samundeswari, R., Nirmala, T., Mahalakshmi, G., Sivassoupramanien, D., Kumar, J. G. S., & Kavimani, S. Conotoxins: Modulators of Ion Channel and their Applications. *International Journal of Pharma Research and Review*, 2014, 3, 57–67.

Saravanan, R., Sambasivam, S., Shanmugam, A., Kumar, D. S., Vanan, T. T., & Nazeer, R. A. Isolation, purification and biochemical characterization of conotoxins from Conus figulinus Linnaeus (1758). *Indian Journal of Biotechnology* 2009, 8, 266–271.

Sarma S. P., Kumar, G. S., Sudarslal, S., Iengar, P., Ramasamy, P., Sikdar, S. K., Krishnan, K. S., & Balaram P. Solution structure of delta-Am2766: a highly hydrophobic delta-conotoxin from *Conus amadis* that inhibits inactivation of neuronal voltage-gated sodium channels. *Chem Biodivers* 2005, 2, 535–56.

Shon, K., Hasson, A., Spira, M. E., Cruz, L. J., Gray, W. R., & Olivera, B. M. Delta.-Conotoxin GmVIA, a Novel Peptide from the Venom of *Conus gloriamaris* Biochemistry 1994, 33, 11420–11425.

Teichert, R. W., Richard, J., Heinrich, T., Doju, Y., & Olivera, B. M. Discovery and characterization of the short kappa A-conotoxins A novel subfamily of excitatory conotoxins. Toxicon, 2007, 49, 318–328.

Terlau, H., & Olivera, B. M. Conus Venoms: A Rich Source of Novel Ion Channel-Targeted Peptides. *Physiological Reviews* 2004, 84, 41–68.

Tucker, J. K., Tenorio, M. J., & Chaney, H. W. A Revision of the Status of Several Conoid Taxa from the Hawaiian Islands: Description of *Darioconus leviteni* n. sp., *Pionoconus striatus oahuensis* n. ssp., and *Harmoniconus paukstisi* n. sp. (Gastropoda, Conidae). In: Severns M., Shells of the Hawaiian Islands – The Sea Shells: 501–514. Conchbooks, Hackenheim. (file://E:/My%20Downloads/2011Hawaiian%20Molluscs-Addendum%20article.pdf)

Usama, A. H. Isolation and characterization of novel conopeptides from *Conus dalli*. MS Disssertation, Florida Atlantic University, 2005.

Violette, A., Leonardi, A., Piquemal, D., Terrat, Y., Biass, D., Dutertre, S., Noguier, F., Ducancel, F., Stöcklin, R., Križaj, I., & Favreau, P. Recruitment of Glycozyl Hydrolase Proteins in a Cone Snail Venomous Arsenal: Further Insights into Biomolecular Features of Conus Venoms. *Mar. Drugs* 2012, 10, 258–280.

Xu, S., Shao, X., Yan, M., Chi, C., Lu, A., & Wang, C. Identification of two novel O_2-conotoxins from *Conus generalis*. *Int. J. Pept. Res. Ther.* 2015, 21, 81–89.

Ye, M., Khoo, K. K., Xu, S., Zhou, M., Boonyalai, N., Perugini, M. A., Shao, X., Chi, C., Galea, C. A., Wang, C., & Norton, R. S. A helical conotoxin from *Conus imperialis* has a novel cysteine framework and defines a new superfamily. *J Biol Chem.* 2012, 287, 14973–14983.

Zamora-Bustillos, R., Aguilar, M. B., Falcón, A., & Heimer de la Cotera, E. P. Identification, by RT-PCR, of four novel T-1-superfamily conotoxins from the vermivorous snail *Conus spurius* from the Gulf of Mexico. *Peptides* 2009, 30, 1396–1404.

INDEX

Printed in the United States
by Baker & Taylor Publisher Services